T0196252

MOUNTAINS, MINERALS, AND ME

MOUNTAINS, MINERALS, AND ME

THIRTEEN YEARS REVEALING EARTH'S MYSTERIES

Albert L. Lamarre

iUniverse

MOUNTAINS, MINERALS, AND ME
THIRTEEN YEARS REVEALING EARTH'S MYSTERIES

iUniverse books may be ordered through booksellers or by contacting:

iUniverse
1663 Liberty Drive
Bloomington, IN 47403
www.iuniverse.com
1-800-Authors (1-800-288-4677)

Because of the dynamic nature of the Internet, any web addresses or links contained in this book may have changed since publication and may no longer be valid. The views expressed in this work are solely those of the author and do not necessarily reflect the views of the publisher, and the publisher hereby disclaims any responsibility for them.

Any people depicted in stock imagery provided by Thinkstock are models, and such images are being used for illustrative purposes only.
Certain stock imagery © Thinkstock.

ISBN: 978-1-4917-7128-0 (sc)
ISBN: 978-1-4917-7129-7 (e)

Library of Congress Control Number: 2015910142

Print information available on the last page.

iUniverse rev. date: 12/14/2015

To my wife, Janet,
with all my love

CONTENTS

PREFACE

Can you imagine being held captive at gunpoint by a Texas rancher's daughter? There I was, out in the middle of nowhere, southeast of El Paso, staring down the barrel of a shotgun pointed at me from just a few feet away. Clad in a dirty white T-shirt and raggedy jeans, she meant business. "What do you think you're doing? This is private property," she said with an intense sneer on her less-than-attractive face, which was surrounded by long, stringy hair. It was obvious she was not to be trifled with.

I was in this predicament because of a simple and honest desire to learn something about a rocky outcrop I could see in the field behind a barbed wire fence. That's what we geologists do, after all. I did survive this ordeal, but this is an example of one of the many escapades I experienced during my thirteen-year exploration career.

My story is that of a minerals exploration geologist, a geologist who goes into the mountains to explore for, among other things, gold, silver, copper, lead, zinc, molybdenum, and tungsten. My career as this minerals exploration geologist was a wonderfully exciting time, complete with fortunate life events that fully satisfied my personal needs. I got to work outdoors in some extraordinarily beautiful terrain, discover and study diverse and exciting geologic settings, and meet many fascinating people. Being an adventurous person, I thoroughly enjoyed working in almost all of the western states in addition to Alaska, Minnesota, Wisconsin, Michigan, and Mexico. My exploration work took me to much of the North American Cordillera, that vast mountain chain that extends from Alaska to Panama.

Having been raised in the small and isolated village of Bath in rural northern New Hampshire, population around five hundred—my identical twin brother and I used to joke that the population is now 498 since he and I left town—I found a career that opened the world to me and provided adventure, intellectual excitement, and mental challenges as I worked in the Rocky Mountains, Alaska Range, the Sierra Nevada, Mojave Desert, Great Basin, Rio Grande Rift, Colorado Plateau, Canadian Shield, and many other places. My geologic credentials were further enhanced by geologic touring of exotic locations in Guatemala, England, and Canada. If it were not for my job, I would never have met and worked with a former circus owner; a miner who routinely went to the bottom of thousand-foot-deep, five-foot-diameter drill holes to enlarge them for atomic bomb placement; and a French doctor living in Guatemala who climbed erupting volcanoes for recreation.

In the pages that follow, I describe some of the places, people, and events from my exploration career as far as my memory allows. More than forty years have moved many of the details to parts of my brain now inaccessible; nevertheless, the details that remain still make a good story. In my narration, I have tried to keep the use of technical terms to a minimum, but a glossary of those necessary terms is included for your reference.

This book will be of interest to anyone who enjoys travel, adventure, science (especially geology), local and regional history, geography, and the excitement of meeting new people and seeing new places. Anyone contemplating becoming an exploration geologist will find this story to be of special interest. So sit back and enjoy some of my additional exploits—having my company office bombed by dynamite-throwing bar patrons in Tucson, being left in absolute darkness in an underground mine in Idaho, and coming face-to-face with a rattlesnake in northern Washington. I hope you enjoy my journey as I vividly recount my first exposure to the geologic wonders of the western United States, the unforgettable characters I met along the way, and the scenic wonders of beautiful landscapes in which I worked. As you travel with me back in time, I hope you resonate as I do with Chris McCandless, the subject of Jon Krakauer's book *Into the*

Wild, who said, "The very basic core of a man's living spirit is his passion for adventure."[1]

I wish to offer my heartfelt thanks to the following people for reading early drafts of this narrative and providing valuable insight and comments: my wife and coconspirator in these adventures, Janet; former coworker and friend Penny Webster-Scholten; and friend Randy Macur.

1. Jon Krakauer, *Into the Wild,* (New York: Anchor Books, 1997).

1. UNDERGRADUATE/ UNDERGROUND

My minerals exploration career began in 1970 when I landed my first job as a geologist during the summer after my junior year at Dartmouth College in Hanover, New Hampshire. I worked for the minerals exploration department of Midwest Oil Corporation, a successful oil and gas exploration and production company that got its start in the "oil patch" near Midwest, Wyoming, and had recently begun a modest minerals exploration program. Many oil companies had entered the hard-rock minerals business and were finding it successful.

The head of Midwest's minerals department at the time, Geoff Snow, is also a Dartmouth College graduate, and one day in the spring of 1970, he called geology Professor Dick Stoiber at Dartmouth and asked, "Do you have a geology student who wants a summer job in Idaho?" When I heard about the opportunity, I immediately said, "Yes!" Here was a chance to see if geology in the field was as fascinating as it was in the classroom. Also, here was an opportunity to explore the world, or at least a small part of it. I had lived nearly all my life in tiny Bath, New Hampshire, and I knew a big, wide world lay out there just waiting to be discovered and explored. Although I didn't know what I wanted to do with my life, I knew it wouldn't be in Bath. Don't get me wrong—Bath was a wonderful community in which to be raised, but the world offered more.

Immediately after finishing my junior year at Dartmouth in June, I flew to Denver to introduce myself at Midwest's company headquarters and then flew to Boise. From there, I rented a car and drove about one hundred miles north on Route 95 to Cambridge, Idaho, my home-to-be for most of the summer and the beginning of my minerals exploration career. When I first

1

met Geoff Snow in Denver, he told me, "You'll like Cambridge, Idaho. When you get back to school, you can tell people you spent the summer in Cambridge. You don't have to tell them it was Cambridge, Idaho, not England." That was my introduction to Geoff's well-developed sense of humor and the beginning of a rewarding personal and professional relationship that continues to this day.

Midwest Oil was conducting exploratory drilling for silver, copper, lead, and zinc at their Cuddy Mountain project in the Seven Devils Mountains near Hells Canyon on the Snake River. Our home base, Cambridge (population around 360), was named for the location of Harvard University, the alma mater of the president of the Pacific and Idaho Northern Railroad who established the town.

This ranching, farming, and logging community was the home of the Office Bar. Midwest's temporary little field office was located adjacent to that bar, and this fine establishment became a frequent hangout for me after work. It was here that I engaged in geologic discussions and debates with my fellow Midwest Oil summer field geologists Dick Kehmeier and Bill Block, both of whom had more experience than I, and with Dave Jonson, the seasoned Midwest project geologist. In conversations that continued late into

Left to right: Driller from Boyles Brothers Drilling Company; Dave Jonson of Midwest Oil, the author; and Dick Kehmeier of Midwest Oil at Cuddy Mountain, Idaho, 1970.

the night at the bar, I heard their stories of working in different geologic environments across the western United States, the challenges they faced, and the scientific discoveries they made. This sounded pretty exciting to me.

While visiting relatives in Idaho in 2009, my wife and I passed through Cambridge, and surprisingly, I discovered the Office Bar was still there. So we stopped for a beer and a glass of wine. The place had not changed—still smoky but welcoming. I was pleased that it remained a place where I could engage in lively conversation as I had forty-four years earlier. Through the window of the bar, we saw the building across the street where I had rented a room above an automobile repair shop while working there those many years before. We watched as cowboys and ranchers walked past the window going about their daily business.

Main Street consists of the Office Bar, the post office, a gas station, and a few small shops, so it was easy back in 1970 to become fully acquainted with Cambridge; its residents readily welcomed me to their town. This was the first western town I became familiar with, and how fun it was that day in 2009 to share my Cuddy Mountain stories with my wife. I reminisced with her about this town being the place where my personal discovery of geologic fieldwork began.

Because I was the youngest person on the Cuddy Mountain exploration crew, I got to ride shotgun in the pickup truck as we went from town to the drill site a few miles away. That meant I rode in the front passenger seat and had to open and close all the cattle gates on the way to and from the property a few miles west of town. It seemed as though there were hundreds of gates—and we *always* closed them behind us.

Working with the two more experienced geologists, who were students at the Colorado School of Mines—Dick Kehmeier and Bill Block—I kept busy splitting and geologically logging drill core, directing the drillers (drilling was done by a drilling company under contract to Midwest), shipping rock samples to the assay office for chemical analysis, mapping the surface geology, and drafting maps and cross sections. The mining industry is a complex, multidisciplinary endeavor that combines elements of science, engineering, business, and finance to locate and mine ore deposits. The industry is fueled by a drive to discover profitable orebodies, and at Cuddy Mountain, the three of us were doing our part to try to make a discovery. Dick, tall and slender as a rail, hailed from western Colorado and was the only westerner among us. Bill, not as tall but broader, was a farm boy from Iowa. Our backgrounds were similar in that each of us had grown up in a small town. Dick and Bill displayed patience with this easterner as they taught me the ropes of how to be a field geologist.

In addition to learning about the technical work required on this project, I learned an especially important lesson about doing geologic work in the field: You can put a can of beans or Dinty Moore beef stew on the engine manifold of the drill rig and have a hot lunch in no time. However, it's important to remember to remove it in time, or you'll have a mess when it explodes on the engine.

Again, since I was the youngest guy on the Cuddy Mountain project, I was assigned to the night shift, responsible for splitting and geologically logging drill core at the drill rig. This was not a bad tour of duty since I was able to marvel at the millions of stars that glistened brightly in that crystal clear Idaho mountain sky. The only light around was from our drill rig, and it was pretty small compared to the heavens above.

Drilling was done with a diamond-impregnated circular drill bit screwed onto the end of cylindrical drill rods; thus it is called diamond drilling. Powered by a diesel engine, the diamonds cut a smooth cylinder of solid rock about two inches in diameter. Once the drillers removed the lengths of rock core from the drill rods, they placed them in trays contained in shallow, three-foot-long, cardboard boxes. They then recorded on each box the exact depth of each length of core.

My job was to spit the core by using a hydraulic ram to break the cylindrical pieces of core in half lengthwise. Typically, one half of each length of core was sent away to an assay office (analytical laboratory) for chemical analysis, and the other half was retained for geologic logging. Geologic logging, my second job, entailed examining each piece of core with a hand lens (magnifying glass) and recording in written format on a clipboard the rock types encountered, alteration, fracturing, mineralization, and anything else that might help unravel the geologic story.

Sometimes geologic logs became quite elaborate, depending upon the complexity of the geology and the creativity of the logging geologist. I've even seen a geologic log done as a poem. I don't recall the poem, but I do remember how creative I thought the logger was. Anyone watching me log drill core would have wondered what on earth I was doing. Minerals in rocks are much more visible when wet, and it was standard practice to spit on the rocks or lick them before examining them through the magnifying hand lens. Splitting and logging core are basic tasks of a minerals exploration geologist, and I had been initiated.

By logging drill core, I was trying to understand what rocks were present at Cuddy Mountain and how they became mineralized (enriched) in copper, silver, lead, and zinc. Patient time spent with the core gave the rocks an opportunity to speak to me and divulge their secrets. I discovered that being a field geologist is like being a detective; the profession requires keen observation skills because not all the clues are obvious—in fact, most of them are not. The profession provides the excitement of a never-ending mystery filled with clues to the complex geologic past of our planet. Planet Earth has experienced volumes of changes in its 4.54 billion years of existence, and the vast majority occurred before we humans arrived. It turns out the geologic past at Cuddy Mountain is represented by folded Paleozoic era (at least 251 million years old) Seven Devils volcanic rocks.

Early in my Dartmouth student days, someone asked me what my major course of study would be. I replied that I wasn't sure yet, but I liked geology and didn't like soft subjects such as sociology because they were not concrete enough for me. When asked why I liked geology, I responded, "Because the rocks don't talk back." In other words, the rocks let me draw my own conclusions through discovery as to what they represent and what story they have to tell. I possess a fondness for that "don't talk back" phrase, and I repeated it many times during my exploration career. Because I enjoyed being outdoors and liked hiking as well, I thought a career as an exploration geologist might suit me nicely. Geologic mapping is just hiking with a purpose, after all. It also didn't hurt that I enjoyed physical labor and didn't mind getting my hands dirty.

Once my professional career began at Cuddy Mountain, I realized, indeed, the rocks don't talk back, but they *do* speak to you. You just have to listen.

By the time I began geologic exploration, old-time pick-and-shovel prospecting had been transformed into the science of exploration. I liked to refer to myself as a glorified prospector who, instead of traveling on foot with a mule, was ferried around in a helicopter and drove fancy four-wheel-drive pickup trucks. My magnifying hand lenses made rock identification easier and more accurate, and sophisticated new techniques for chemical analysis enabled identification of lower concentrations of metals and minerals. Before going into the field, I would don my trusty, wide-brimmed hat for sun protection, put on my canvas Filson field vest that held much of my field gear, put around my neck the leather cord that held my hand lenses, ensure my backpack contained water, lunch, and plastic or canvas sample bags, secure my topographic maps and aerial photographs in a metal clipboard, tightly cinch up my field boots, and, most importantly, be sure I had my dependable rock hammer. A geologist is never without his hammer.

Minerals exploration is a fascinating, scientific research activity. As someone once said, it is through science that Mother Nature reveals her mysteries to us. There are lots of mysteries involved in minerals exploration. To unravel those mysteries and discover an ore deposit, you must first have an idea of where a mineral deposit might be found or be able to visualize what a deposit might look like. This leads to a geologic or conceptual model that you can test through data gathering and interpretation. Sampling, mapping, geochemical and geophysical surveying, and drilling are done during this stage. Each phase of hypothesis testing may require years of costly work, and most exploration projects do not yield economically viable mineral discoveries, called ore deposits or orebodies. Even though an exploration geologist may discover minerals, their concentrations must be high enough that they can be mined at a profit; most discoveries do not meet this critical criterion.

From initial mineral discovery to mine production can be five to fifteen years or more. Despite improvements in science and technology, the investment in exploration remains a high-risk proposition. As someone once quipped, the modern geologist sells hope, backed up by some fancy colored maps. The promise of large financial rewards is what stimulates the necessary commitment and level of investment required by a company in this industry. For an exploration geologist, the possibility of someone pointing you out as, "A person who discovered an ore deposit," is the ultimate compliment and provides the stimulation to pursue this low-rate-of-success business. I personally know only three geologists who have an ore discovery to their credit, even though I have worked with more than a hundred geologists.

Western Idaho around Cuddy Mountain is beautiful country. Although there are no soaring mountain peaks like in central Idaho, the broad, green meadows with grazing beef cattle and surrounding evergreen-tree-covered hills are peaceful and serene. McCall, Council, New Meadows, and Payette are small, friendly towns surrounded by lakes and forests. I especially enjoyed seeing the logging trucks coming out of the forests with their heaping loads of logs that reminded me of New Hampshire's logging industry.

Cuddy Mountain wasn't all work. One Sunday, Bill Block and I floated the Weiser River in inner tubes, a slow and leisurely float with beer cans in hand as we luxuriated in the serenity of a lazy journey downstream. I also spent one weekend in the Wallowa Mountains across the Snake River in Oregon where a cousin of my father's lived on a ranch. She was thrilled at my arrival because I was one of the few East Coast visitors she had received. I had never met her before, and it was fun to listen to her stories of growing up in New Hampshire and then moving west.

As mentioned previously, Hells Canyon borders the Cuddy Mountain project site on the west and is the deepest gorge in North America. Carved into the Columbia River basalts by the Snake River, this north-south canyon defines the border between Oregon and Idaho. Hells Canyon was significant to me because it provided my first opportunity to examine geology that was fully exposed over a large vertical thickness—7,913 feet, to be exact, from canyon rim down to the water. The abundant vegetation and soil that cover the rocks in New England prohibit this kind of examination back home. So imagine how excited I was to see in the walls of Hells Canyon a thick sequence of distinct, black, basalt lava flows piled one on top of another. Volcanoes had once been active nearby.

At the northern (downstream) end of Hells Canyon, the Snake River becomes substantially larger after the Salmon River flows into it near Lewiston, Idaho. Barges travel the 465 miles from the Pacific Ocean up the Columbia River and then along the Snake River to Lewiston where they take on wheat and other grain crops. Consequently, Lewiston is called "Idaho's Seaport."

Upon completion of drilling late in the summer, I said good-bye to the town of Cambridge and was sent to work at the Ima mine, another Midwest Oil project, this time on the opposite side of the state, almost in Montana. Disposition of the Cuddy Mountain project would have to wait for data evaluation to be done during the winter. At the inactive Ima tungsten mine near Patterson, Idaho, just a wide spot on the county road east of Challis, Midwest was evaluating the bulk mining potential there. Located in the Lemhi Range, the Ima mine is at the northern boundary of the Basin and Range physiographic province that extends from southern Idaho through Nevada, Utah, California, and Arizona to New Mexico. The province is named for its parallel, elongate mountain ranges and intervening dry valleys or basins.

While working at the Ima mine, I stayed with a nearby ranching family where I witnessed the day-to-day struggles and pleasures of ranch life. I especially enjoyed the delicious and hardy homemade meals they shared with me. This family also provided my first horseback-riding experience on their Pahsimeroi Valley ranch. The ranch was near where Sacajawea, Lewis and Clark's Lemhi-Shoshone interpreter, was born. As I rode around the ranch, I tried to imagine what it must have been like to live here in the late 1700s as she did, on sagebrush-covered plains surrounded by forested mountains. On their expedition to find the Northwest Passage, Lewis and Clark crossed over the Continental Divide at nearby Lemhi Pass on August 12, 1805. I wondered if they had walked and ridden their horses across the tungsten-bearing quartz veins that later became the Ima mine. Even though out of curiosity they may have bent over to pick up a piece of the gleaming, white quartz with its reddish-brown tungsten mineral, it would not have had any value to them because it had no use at the time.

The Ima mine project provided my first experience going into an underground mine, and because it was entirely new and foreign to me, I found working underground to be fascinating and invigorating. I had never even been in a cave before. What a thrill it was to be hundreds of feet below the ground surface entirely surrounded by rocks. I loved it!

But there is nothing darker than being deep underground without a light. I learned that in dramatic and scary fashion at the Ima mine. Proper lighting while working underground is normally achieved by a lamp (carbide or battery powered) affixed to a hardhat—additional emergency light sources are recommended. One day Geoff Snow and Don Adair, another Midwest geologist with whom Geoff had gone to graduate school at the University of Utah, lowered themselves by rope down an ore pass from deep within the mine. My job was to stay behind at their point of departure and make sure their ropes held tight.

Suddenly my carbide light went out, and I had never known such darkness. I became frightened and fought off panic. An eerie black void surrounded me, and I was alone, needing to see what I was doing. Geoff and Don were depending on me for their safe return. A carbide lamp works by allowing water to drip into a container of carbide powder; a chemical reaction generates acetylene gas, which is then burned to produce a flame. Why we had these old-fashioned carbide lamps rather than battery-powered lights is still a mystery to me. Fortunately, I had some extra water and carbide, but in the darkness, the task of refilling and relighting the lamp would have to be done entirely by feel. I dreaded the possibility of dropping either of the vital ingredients. How would I ever locate them in such total blackness without losing my balance and disappearing down the steep ore pass myself? Slowly and meticulously, I replenished the water and carbide in the lamp and relit it. Whew! I was successful in regenerating light, and thankfully, my two partners emerged from below as scheduled. If this was a test for a young geologist, I guess I passed.

This was a good example of the value of my trusty Filson field vest, which was loaded with pockets to carry all my stuff, and fortunately, one pocket held matches that were readily accessible to light the carbide lamp. The vest's sturdy canvas construction sported additional pockets for carrying other important items like colored pencils and a pencil sharpener, felt-tipped marking pens, altimeter, field notebook, Swiss Army knife, and sometimes even an acid bottle. That was the only downside to the vest's canvas construction. Carbonate minerals like calcite and rocks containing calcite—such as limestone and marble—effervesce or release carbon dioxide gas when they come in contact with acid. The resulting fizzing or bubbling action allows the geologist to quickly identify carbonate minerals in rocks. The only problem, however, is that bottles of acid invariably leak around their caps, so after a while, the acid-bottle-containing pocket of your vest develops ratty-looking holes.

Silver had been discovered here around 1900 in large, white quartz veins (a vein is a sheet-like body of minerals that has precipitated from hydrothermal solutions [hot water] into cracks and crevasses of rocks), but by 1937, tungsten became the primary metal of interest. In 1938, the Ima mine boasted being the second largest tungsten mine in the United States. My job was to collect samples from the walls of the old tunnels and adits of the underground mine (a tunnel is a horizontal excavation dug into a hillside with an entrance at each end; an adit is a horizontal excavation dug partway into a hillside with only one entrance). In addition to collecting samples for chemical analysis to determine the tungsten content, I soon accumulated beautiful and rare specimens of reddish-brown huebnerite, the tungsten-oxide ore mineral mined here until 1957 when the mine closed. Tungsten's value lies in its use as a hardener in steel.

About six years after working at the Ima mine, I returned there to collect more specimens for my growing personal rock collection. Using my rock hammer, I broke off large pieces of rock from the walls of the mine workings (the tunnels and adits), examined them for the quality and quantity of the tungsten-bearing huebnerite mineral, and then further reduced the best specimens to fist-sized samples. These I put into plastic or canvas sample bags that I immediately labeled so as not to confuse the samples. The difficult part was carrying them out in my backpack. Breaking and collecting rocks is hard work (prisoners used to do this, didn't they?), but for me it was a labor of love.

The return of exploration companies to old mine sites like the Ima mine was a common exploration practice at the time. Part of the reason was that higher metals prices may have improved the economic potential of a closed mine. Or, just as commonly, the old-timers probably mined the high-grade ore but left the lower-grade material behind. With more modern techniques available for mass-removal mining, this low-grade material could become ore. Whereas the mid-twentieth century miners at the Ima mine were interested only in the one- to ten-foot-thick quartz veins and the tungsten minerals in them, we were exploring the possibility of finding tungsten disseminated within the rocks away from the veins. As part of my job, over the years I went to many old, abandoned mines such as this. Alas, the Ima mine did not pass muster, and Midwest did not put it into production.

Copper Gulch at Grimes Pass northeast of Boise, Idaho, was another Midwest Oil project where I spent a short time still later that summer collecting samples of the stockwork of quartz-copper-molybdenum veins. A stockwork is a network of crisscrossing veins. If the density of the veins comprising the stockwork is high and if the copper and molybdenum content is also high, then this could constitute ore. This property was located in the Boise Basin near the small logging and recreation towns of Crouch and Garden Valley. The South Fork of the Payette River flows past these towns, and I envied the groups of people testing their skill at rafting its class III and IV rapids. But I was too busy learning about the geologic setting to have time to partake. However, many, many years later, my wife, Janet, and I had the opportunity to raft the Payette River, and we learned for ourselves how much fun those people were having and how much skill is required to successfully float those rapids.

As my summer with Midwest was winding down, I worked in Little Cottonwood Canyon in the Wasatch Range northeast of Salt Lake City, Utah. I spent only a short time there on this molybdenum exploration project, but what I distinctly remember were the beautiful, large (inches long) potassium-feldspar crystals in the granitic rocks exposed in the canyon walls. The incredible mountain scenery that surrounded me was memorable also; beautiful Alta ski area was just up the road.

From core splitting to sampling underground, my first time out west afforded me an education unavailable in a classroom, not only in the diverse aspects of geology but also in cultural awareness and western recreation. I had learned an incredible amount about geology and what it takes to be an exploration geologist, and I liked it. On my flight to Denver at the beginning of the summer, I had been filled with excitement about what the summer would bring. I had not been disappointed. Maybe I had found my place in the world. I vowed to return to Midwest Oil Corporation the following summer for additional experience and adventure.

2. NORTH TO ALASKA

During my senior year at Dartmouth, my geology classes held even more excitement for me because of the previous summer's relevant fieldwork. I was now sure that I was on the right track with geology as my major course of study. I spent the school year in anticipation of getting back into the field, and after graduating from Dartmouth in June 1971 with a bachelor's degree in earth sciences, I eagerly returned to Midwest Oil Corporation.

I was thrilled to be informed by Geoff Snow that this summer I would work on a geologic reconnaissance crew in the massive Alaska Range between Fairbanks and Anchorage. I flew into Anchorage in late June with Bruce Bouley, another Midwest Oil summer hand, who was a geology student from Wesleyan University in Connecticut; we spent the night at what to me was the ritzy Captain Cook Hotel. This hotel was multistoried and fancier than any hotel I had ever been in. Bruce had no more exploration experience than I did, so we eagerly shared our excitement about getting to explore the Alaskan wilderness. We also shared some anxiety—would we be up to the task? Today's website for the Hotel Captain Cook states that this hotel "… provides the downtown launching pad for charting your Alaskan adventure," complete with the question, "Are you ready to start?" Perhaps this is exactly what happened to me—my adventurous geologic career was launched right there at the Captain Cook Hotel.

The next day, Bruce and I flew to Fairbanks in the near center of the state and then drove south into the Alaska Range to our first base camp at the town of Healy on the Alaska Railroad, where we met other Midwest geologists who were already at work. There isn't much in Healy besides a few old buildings and the Otto Lake Lodge—a nondescript motel—that served as our headquarters just beyond the northeast corner of Mt. McKinley National Park. (In 1980, the park was enlarged threefold to six million acres, and its name was changed to Denali National Park & Preserve. I will refer to it as "McKinley" since that is what it was called when I was there.)

For those of you who have read the book or seen the movie *Into the Wild*, Healy is where true-life Chris McCandless started his fateful journey into the wild Alaskan wilderness. The most I remember about Healy is that the Nenana River flows through Healy Canyon south of town, and the water was an ugly, cement-gray color from all the glacial silt it was transporting.

This minerals exploration program was carried out with the aid of a bright red, two-passenger Bell G3B helicopter with a Plexiglas enclosure that provided near-perfect viewing of stupendous scenes below us. Working as a two-man crew, Bruce and I became a team; after all, we did look alike, both being relatively short and blond-haired, and both were from New England. Bruce was an intelligent and engaging guy with a twinkle in his eye and a good sense of humor; we got along well. I especially appreciated that mischievous look he always had. It made me wonder what he was up to. We subsequently ran into each other over the years at various geologic events and became lifelong friends.

George Moerlein, a seasoned Dartmouth graduate who lived in Anchorage, was a consulting geologist in charge of Midwest's Alaska operations for the summer. He was a bearded curmudgeon who tolerated no nonsense from his summer hands. Dave Jonson, Steve Zahony, and Noel McAnulty, Jr., all full-time Midwest Oil geologists, formed the heart of the field crew. Toni Hinderman, also from Anchorage, was a consulting geologist who was added to the crew because of his extensive work experience in the state.

Troy Cook, an experienced flier from the Viet Nam War, was the helicopter pilot. *Experienced* is the important word here. That meant Troy had lots of flying time under his belt and was conscientious about flying safely. Helicopter flying is inherently dangerous, but we were always careful and safety conscious. "*Never, ever* stand up straight while entering or exiting the helicopter" was continuously drummed into us. Those rotating blades over your head are unforgiving. And you *never, ever* walk to the rear of the helicopter where the tail rotor is spinning.

In spite of this training, we did do some nerve-racking things. Troy dropped us off each morning at the highest peak around, and because of the steep and rugged terrain, sometimes there was no level spot on which to land. In that case, Troy would set one skid of the helicopter on a large boulder, with the other skid hovering in the air, and shout, "Jump!" Hearing his command, we'd leap out of the helicopter onto the boulder-strewn ground. Fortunately, no mishaps occurred during this practice even though there were many such teetery landings. You can imagine the adrenaline rush the first time I did this. "Troy wants me to do what?" I had never been in a helicopter before, to say nothing about jumping out of one!

To cover more terrain, Bruce and I would split up and walk downhill all day, examining the geology, taking field notes, and collecting rock samples. Sometimes I even took time to admire the scenery. That summer, we hiked many miles over sharp and narrow ridges, across glaciers, and through too many boulder fields to count. Troy picked us up at the end of each day, but sometimes by then the air had warmed so much that there wasn't as much lift for the helicopter rotor blades as there had been in the morning. To take off, we would get some assistance from gravity. By revving the engine, Troy could get us a little off the ground, and then he would let the helicopter fall sideways off a cliff or slope. We would fall fast, and the speed of the fall would give Troy enough airspeed to steer. As we reached the thicker, lower-elevation air, we would regain the lift we needed.

This approach was always exciting, to say the least, but I never really did get used to it. I was never frightened, but I got close to being so when we would fly level over flat areas and then abruptly come across a valley. As the ground suddenly dropped from beneath me, my stomach seemed to drop also. But I must have been a daring sort because I loved the feeling as we lifted off from the ground and 360 degrees of scenery came into view.

A helicopter view of a glacier in Sheep
River Valley, Alaska, 1971.

Bruce and I carried radios so we could communicate with
each other and with Troy to tell him when and where to pick us
up at the end of each day. We also carried flares to shoot into the
sky in the event the radios failed or if cloud cover hindered Troy's
ability to see us. I had to use my flares one day and was relieved
to know they really worked as I watched their bright illumination
travel high into the sky. Low clouds were obscuring Troy's ability
to see me, but he did see the flare and safely picked me up.

Hearing the chatter of the helicopter's rotor blades as Troy
approached was always comforting and reassuring. Even more
satisfying was to be sitting in the helicopter at the end of a long
day, heading back to base camp. Those days of carrying backpacks
heavily ladened with rock samples put a big strain on my knees,
and I sometimes feel the consequences today with aches and
pains. When back at base camp, I would eat a hearty meal, discuss
with my co-explorers what we had each seen that day, and then
settle down to plot on maps the findings of the day's traverses.

During that summer, we worked our way south from Healy
to Anchorage, staying at base camps in Cantwell, Stephan Lake,

and Talkeetna, all the time in remote and beautiful country of interior Alaska. Cantwell lies near the Nenana River immediately adjacent to Mt. McKinley National Park and has a population of only about two hundred hardy souls. The sky stayed light until late at night because we were so far north; I remember reading a newspaper outside at ten o'clock.

Stephan Lake Lodge, the next base camp, is accessible only by air, and every night Troy parked the helicopter right next to the lodge while float planes parked on the lake at the dock. I recall playing lots of cards one week when we were grounded because of clouds being right down on the deck that prevented flying. We were all frustrated and eager to get back into the field. One day when we did fly, I saw below me hundreds of bright red salmon that had swum up the creek to spawn. What a sight. I had never seen that before. Five-mile-long Stephan Lake is recognized as a prime spot for lake and stream fishing in interior Alaska, but I don't remember ever having the chance to try my luck. Even though I have a photograph of me in waders, I can't imagine that George Moerlein ever gave us time off to go fishing. He was very much a git-r-done kind of person, and we worked every day that weather permitted; otherwise, there was no time off.

Talkeetna, our last base camp, is the favored jumping-off point for mountaineers attempting to climb Mt. McKinley, the highest point on the North American continent. About five hundred people live in Talkeetna, and, being only sixty miles southeast of the grand mountain, they enjoy one of the finest views of this 20,320-foot-tall peak. Talkeetna, some say, served as the inspiration for the 1980s television series *Northern Exposure*, but I can't verify that. The small airstrip at Talkeetna always seemed to be busy with bush pilots ferrying climbers to and from their base camps set up on glaciers rimming Mt. McKinley. Twenty years later, I worked with a person at Lawrence Livermore National Laboratory who had flown into and out of Talkeetna, but he does not have the same fond memories I do. In an attempt to scale Mt. McKinley, he experienced severely frostbitten feet, so much so that some toes were amputated.

I saw lots of grizzly bears that summer; in fact, we all wore a .44 Magnum revolver on our hip for bear protection, though none of us ever had to use it. That's a good thing since I had never fired a revolver before that summer. However, in preparation for possibly having to shoot in self-defense, I took target practice one of the first days I was in Alaska. I never became skilled in the use of that revolver, and I didn't like the recoil and muzzle flash when I fired, but at least I became a somewhat competent shooter. Each revolver held six bullets, and we joked that five were for the bear; if they didn't stop the bear from charging, the sixth one was for the shooter.

Watching the grizzlies from the air as they raced away from the howling helicopter was always exciting. For being such bulky animals, they certainly run fast. Fortunately, I made enough noise while hiking that it must have scared the bears away; I never did see one from the ground. I probably could not have outrun one. Moose and caribou were plentiful also, but they did not present the same danger that grizzly bears did. One day I came across a moose feeding in a meadow, and we both stared at each other from about fifty feet away. I don't know who was more curious, the moose or me!

The physical setting of this Alaskan wilderness was as amazing as the fauna. We commonly flew low enough over broad glaciers that we could see the blue ice deep in their crevasses. I hiked over ice fields, glacial moraines, and rock glaciers, and along ridges, cirques (heads of glacial valleys with amphitheater-shaped basins carved by glacial plucking), and arêtes (knife-edge ridges formed where two or more cirques were eroded back-to-back by the glaciers). Virtually every glacial feature that one learns about in earth science class was superbly represented.

Most of each day I spent above tree line with vistas that seemed to stretch forever—my kind of country. When lunchtime arrived, I always tried to find a place to stop that had a spectacular view where I could appreciate the exhilaration of high altitudes that I was experiencing at these high latitudes. As I hiked along during my day's traverse, I usually could see no evidence of humans. I truly felt like a frontiersman. It's no wonder Alaska has seventeen national parks. Back at Dartmouth, I had done a lot of backpacking in New Hampshire's White Mountains, but nothing there matched the grandeur I was experiencing in Alaska.

Although our exploration work was performed outside the national park boundaries, the most impressive day of the summer was seeing the summit of Mt. McKinley for the first time. At 20,230 feet high, it is a magnificent and enormous mountain—the centerpiece of the Alaska Range. The peak had been clouded in most of the time I had been near it, but when the weather cleared one day, I finally saw its perpetually snow-covered summit—breath taking! Mt. McKinley stood much higher in the sky than I had ever imagined. No wonder it's the namesake of Mt. McKinley National Park. Trite as it is to say, I felt so small in comparison to this mountain. I realized that geology has a profound impact on how we humans view the world around us. We are only a small part of this world.

We have the geologic principle of plate tectonics to thank for the majesty of Mt. McKinley. The Pacific tectonic plate—a tectonic plate is a segment of the Earth's crust that moves in response to forces in the underlying mantle—is being thrust under the North American plate along a linear trench (called a subduction zone) parallel to Alaska's southern coast. As the Pacific plate moves north and plunges beneath Alaska, the North American plate is crumpled and lifted into the soaring peaks of the Alaska Range, most notably Mt. McKinley. The magnitude 9.2 Alaska earthquake of Good Friday, March 27, 1964, is clear evidence that plate tectonics remains active up there.

This Great Alaska Earthquake caused 131 deaths as a result of ground fissuring, collapsed buildings, and tsunamis (so-called tidal waves), some as high as seventy feet. Interestingly enough, one of those tsunamis reached Crescent City in northern California, where twelve people died when a wave of water twenty-one feet high came ashore. I've been to Crescent City, and I can see how this could happen; the town sits right on the ocean at sea level. This destructive tsunami is the largest one to have struck California

Typical Alaska scenery that I constantly
marveled at from the helicopter, 1971.

in recorded history, although north-coast American Indian legends
tell of giant waves that caused much destruction in prehistoric
time.

Can you imagine withstanding approximately four and a
half minutes of violent shaking of the ground beneath your feet?
An aunt and uncle of mine living in Anchorage at the time did
experience that, along with their four young children. Fortunately,
they were not seriously injured, but their arms were bruised from
bouncing off the walls. And they certainly were scared. With a
magnitude of 9.2, this is the largest earthquake ever felt in North
America and the second most powerful in the recorded history
of the entire world. (A magnitude 9.5 earthquake struck Chile
in 1960.)

This Alaskan earthquake was so strong its effects were felt
three thousand miles away in Texas where water sloshed out of
swimming pools. Vertical displacement (movement) along the
fault that ruptured and produced the earthquake was measured at
thirty-eight feet, which is a huge amount for a single earthquake.
Over many millions of years, tens of thousands of earthquakes

comparable to this in magnitude, and smaller, were required to push Mt. McKinley to its towering 20,230-foot elevation. You can think of earthquakes as the incremental steps of mountain building.

Most people (including some geologists) find it difficult to comprehend the amount of time required to accomplish such geologic feats as mountain building. Two hundred million years were required to create the entire North American Cordillera (the mountain chain along North America's west coast), including Mt. McKinley. This sounds like a long time, doesn't it? However, it comprises only 4 percent of all geologic time since planet Earth is 4.6 billion years old. This length of time measured in billions of years is what geologists refer to as Deep Time.

To understand Deep Time, I cannot do a better job describing it than Jennifer H. Carey does in her book, *What's So Great About Granite?* She says, "Scientists estimate the Earth is about 4.6 billion years old—4,600,000,000 years. That number can be hard to fathom. If you compare the age of Earth with a single [calendar] year, with Earth forming on New Year's Day, then the dinosaurs went extinct on December 25 [of that year] and our species, Homo sapiens, came to the New Year's Eve party around 11:00 p.m. Recorded human history fits within the last minute of the year, but geologic history spans the entire calendar year."[1] After digesting her description, you must admit, geologic time is a fascinating subject.

Geologists routinely use the geologic timescale to specify the ages of rocks and geologic events, but this timescale is mind-boggling to most people. It may sound like a simple statement to say that dinosaurs died out sixty-five million years ago, but for a person who cannot imagine what his/her hometown might have looked like just 650 years ago, thinking in terms of millions of years requires a paradigm shift; one must be capable of temporal thinking. Deep Time is one of the most difficult concepts for the

---

1.   Jennifer H. Carey, *What's So Great About Granite* (Missoula: Mountain Press Publishing Company, 2009).

layperson and geologist alike to grapple with, myself included. The appearance of humans on the planet at 11:00 p.m. on New Year's Eve, as Ms. Carey says, gives a sense of antiquity to our Earth as well as humbleness to mankind. Humans have been on this planet Earth for only about 0.004 percent of its history. By the end of my exciting summer in Alaska, it was clear to me that geology affects our everyday lives, and I wanted to be part of unraveling the planet's geologic story.

At the end of fieldwork in September, I flew in the helicopter with George Moerlein to his home in Anchorage where Troy put us down on George's front lawn. You don't get to do that just anywhere! I remember telling my friends later that this job had been a scam—I got paid for doing all summer what other people paid good money to do during only two weeks of vacation! Indeed, I had many stories to tell when I arrived at graduate school that fall at the University of Wyoming in Laramie. During my senior year at Dartmouth, I had realized there was much I still needed to learn about geology, so graduate school was the next step in preparing for my career. The University of Wyoming is where I decided to go.

My summer in Alaska turned out to be one of those turning points in life that I expect everyone experiences at one time or another. That summer's experience convinced me that exploration geology was something that I was good at and enjoyed, and was the direction my life should take.

3. TAKE A WRONG TURN,
SO RIDE THE BULLET

I entered the University of Wyoming's geology graduate program in September 1971 fully charged, excited, and ready to learn. My classes included ore deposits, mineralogy, and structural geology, but unfortunately, I found them lacking in stimulation—just more of the same classroom work I had done the previous four years. Maybe I was burned out on school and needed some time off, I thought. What clinched it for me was when the ore deposits geology professor at the university mispronounced the name of one of the major copper mines in Arizona. The mine was world-renowned for its thick blanket of high-grade ore, and I thought a distinguished professor surely should know the correct pronunciation of this mine's name. That was juvenile thinking, I know, but that happens sometimes when you're twenty-two years old. So in February 1972, after a little more than a semester, I acknowledged to myself that I had taken an educational wrong turn, and I left graduate school to go back to work for Geoff Snow at Midwest Oil Corporation in Denver (to be strictly correct, the office was in Lakewood, a Denver suburb).

Thank heavens Geoff would have me back. I had found that he was always eager to share his knowledge with others and teach new geologists the tricks of the trade. Although short of physical stature, he was a dynamic and inspirational leader with an outgoing and welcoming personality. So I was thrilled when he agreed to have me come back to work for him, and over the years he became my mentor. This was one of the luckiest things that ever happened to me.

Under the direction of Geoff, Dave Jonson, and Steve Zahony, I worked in the Midwest Oil office located in the Colorado

Rockies foothills where I enjoyed tremendous views eastward to the metropolitan Denver area. That spring I kept busy compiling geologic information, reviewing geologic maps, and plotting data from various Midwest projects: Ima mine, Copper Gulch, and a Wyoming massive sulfide program. I did get out of the office once when I flew to Boise, Idaho, and then drove an hour and a half north to the little town of Crouch to show the Copper Gulch property to a Longyear Drilling Company representative. We were soliciting their bid for a drilling program that Midwest would begin when the snow melted. Normally, drilling was the biggest expense in exploring a property, so we always wanted the best bid we could negotiate.

Late in the spring, Geoff assigned me to work on an exploration program that Midwest was preparing to conduct in partnership with General Crude Oil, an oil company based in Houston. Called the North Central Minerals Venture, this partnership's goal was to discover copper, lead, and zinc deposits in Wisconsin, Michigan, and Minnesota. In preparation for my role, in May I flew to Michigan Technological University in Houghton, Michigan, to attend the annual Institute on Lake Superior Geology and learn about the regional geology. This was my introduction to Great Lakes geology, and I soaked up all the information I could.

However, when I arrived there, I thought I was at the end of the Earth way up there at the northern part of Michigan's Keweenaw Peninsula where it juts out into Lake Superior in the shape of a shark's fin. The snowbanks were still piled high in May.

Michigan Tech was established in 1885 as the Michigan Mining School for training mining engineers to operate the copper mines that had been developed on the peninsula starting in the 1850s. The Upper Peninsula of Michigan became a famous copper mining region, and until the 1920s it produced 90 percent of the copper used by the United States.

In further preparation for my upcoming work in the Great Lakes area, I flew to Houston to meet with General Crude Oil's geologists. We discussed plans for the summer and agreed upon an exploration strategy. That flight to Houston was a real treat because I flew on Midwest Oil's corporate jet. The pilots of this Sabreliner jet had affectionately and appropriately nicknamed it the Bullet. Imagine my thrill as a young twenty-three-year-old to be flying across the country on a private jet! I felt important! While in Houston, I also attended my first professional baseball game. My sheltered life was being transformed rapidly: meeting corporate executives, flying on a private jet, and attending a professional sports event. I was becoming more worldly-wise and self-confident as my horizons expanded.

From Houston, I then flew—again on the Bullet—to Ottawa, Ontario, to meet with geologists from the highly regarded Geological Survey of Canada (GSC). Because the geology on the United States' side of the Great Lakes is the same as that across the border in Canada, and since GSC geologists had spent a lot of effort mapping on their side of the border, I wanted to learn what they knew about the geology. Indeed, they saved me a lot of time by sharing their vast knowledge and experience gained while working on the Canadian Shield.

On this trip, I met Bob Hodder for the first time, a geology professor at the University of Western Ontario who was also a consulting geologist for Midwest. He joined me in the meeting with the GSC geologists. Having spent many years as an exploration geologist himself, Hodder was now teaching exploration geology at one of Canada's premier universities. One year later, he would become my master's thesis advisor and another valued mentor.

In June, I started fieldwork on the North Central Minerals Venture project. Having rented a room in a house in Edina, a suburb of Minneapolis, I made many reconnaissance trips to Wisconsin, Michigan, and northern Minnesota from that base of operations. I conducted much of my work around the towns of Rhinelander, Wausau, and Eau Claire in northern Wisconsin, where active base-metal (copper-lead-zinc) mines exploited the Thornapple and Ladysmith massive sulfide deposits. These deposits served as our exploration model.

Working in terrain so close to active mines was exciting, and the rocks were older than any I had seen before. These ancient

Precambrian rocks (2.6–2.7 billion years old) of the Superior Province had been scraped clean by ice age glaciers and then covered with up to one hundred feet of glacial debris called glacial drift or till. As a consequence, rock exposures are rare. I felt right at home while walking along stream channels looking for outcrops and searching out quarries that might expose bedrock; these were the same conditions as in my home state of New Hampshire. The prevalence of familiar dairy farms was comforting also. Having been a long-time 4-H member while growing up, I felt right at home taking notes with my pen that had a 4-H symbol on it.

Because of the presence of this glacial cover material, geophysical techniques are the primary exploration tools used in this environment to "see" through the glacial till. These tools send a pulse of electromagnetic energy into the Earth that is reflected and picked up by geophysical sensors. Usually, the energy-sending and -receiving devices are carried by an airplane or helicopter. We contracted with Geoterrex, an Ontario-based geophysics company, to fly in a grid pattern close to the ground while towing a geophysical instrument (called a bird) beneath the airplane. Geoterrex used a lumbering PBY Catalina, also called a Widgeon, a twin-engine propeller-driven plane with its wing mounted above the fuselage. This impressive "flying boat" was a workhorse during World War II when it flew on antisubmarine patrols; it was now a workhorse in the exploration business. After flying the designated pattern, a Geoterrex geophysicist interpreted the electromagnetic signals that were received. I was not able to evaluate the results since I had not been trained in this specialized field. I didn't get to ride in the plane either!

Unfortunately, after conducting geophysical surveys on a number of different tracts of land, we had no favorable geophysical responses. So at the end of August, I returned to Denver after a geologically unproductive summer. I had not uncovered any exploration targets that warranted follow-up work. However, ruling out the need for further exploration was itself important because no additional financial resources would be spent there. To demonstrate how low the chances of success here were, in 1975 the discovery of the Crandon deposit in nearby Forest County was announced by a competitor—it had been found after twenty-four other properties in the area had been drilled and found wanting. This summer had not been nearly as exciting as my previous summer in Alaska, but summering in Alaska was pretty hard to beat.

My time spent around the Great Lakes did have educational value, however, because I expanded my knowledge of a geologic environment that contained other important ore deposits in addition to the copper-lead-zinc deposits we were in search of. In northern Minnesota, I toured open-pit hematite and taconite iron mines in the Iron Range near Hibbing and Bemidji. I also went underground in an active iron mine at Iron Mountain in Michigan's Upper Peninsula. Most iron ore produced in the United States comes from huge open-pit mines in these two states.

The rocks containing the hematite, which is high-grade iron ore (containing more than 50 percent iron), and taconite, which is low-grade siliceous iron ore (30–40 percent iron), consist of bands of lithified (meaning turned into rock) fine-grained sediments that were deposited in oxygen-rich Precambrian seas (1.8–2.5 billion years ago). This banded iron formation, as it is called, is a beautiful layering of alternating and swirling bands of gray to black to reddish-black iron-oxide minerals and chert (silica). The rocks are magnetic and gorgeous; when cut and polished, they make impressive bookends. I have a set myself. These rocks were recognized for their economic value as far back as the mid-1800s.

While touring these iron ore mines, I was close to Lake Superior, the largest of the five Great Lakes. It's hard to believe these huge bodies of water, containing 20 percent of the world's fresh water, were carved out by the Pleistocene continental ice cap about twelve thousand years ago and then filled with glacial meltwater. The ice sheet was up to two miles thick and incredibly powerful.

I still have fond memories of riding a bicycle around many of the numerous lakes that comprise the Minneapolis metropolitan area. Not knowing anyone in the area, this was an enjoyable way to spend evenings after work.

After returning to Denver, I was immediately sent off to my next project, exploration of the Florida Mountain silver property in the Owyhee Mountains of southwestern Idaho, close to Oregon. George Ambariantz of Earth Resources Company in Golden, Colorado—Midwest's joint venture partner in this effort—was the project leader. Because of the high cost of exploration and low chances of success, it was common for mining companies to join efforts in exploring a mineral property. George was a seasoned geologist whose family came from Armenia. I had never met anyone whose relatives were from the Black Sea area, so I talked a lot with him about his heritage. There was more to learn on this project than just geology.

We stayed at the historic Idaho Hotel in Silver City, Idaho, an old mining town dating from the 1860s that was almost a ghost town in the 1970s. A hundred years earlier, this was a thriving mining district that produced thousands of ounces of gold and silver. But in the fall of 1972, George and I and our drillers were about the only people in town.

We contracted with a local cat-skinner (bulldozer operator) to build drill-rig access roads and excavate trenches high up on the side of Florida Mountain with a huge D9 Caterpillar bulldozer. This bulldozer was so big it sounded like a freight train moving along the hillside. By digging through the knee-high sagebrush and soil and thus exposing the underlying bedrock, George and I were able to collect pickup-truckloads of rock samples for chemical analysis. Once the results were returned from the analytical laboratory, we drew contours of the silver values on a hand-drawn map of the trenches to identify the higher-grade areas. Now came the fun part: based on the distribution of the higher-grade areas on the surface, we decided where to drill exploratory holes to see if there was silver ore at depth.

We didn't spend much money on drill road construction, so the roads were merely single-purpose tracks across the rocky hillside; they were always rough and often barely passable. They were not built for speed, and one time as I was driving along—much faster than I should have been—I almost hit my head on the pickup-truck roof as I went over a bump. I got tossed around more than a softball at a church picnic.

Ekland Drilling Company of Elko, Nevada, did the drilling for us, and I collected and geologically logged the drill samples. We drilled the property during October and November, and many days were spent fighting the weather. It seemed we constantly had to deal with frozen equipment and drill rigs stuck in the snow. One of my notes from that time period records this comment, "It rained like hell all day!" I remember several instances of mapping the geology in the rain and snow. In general, however, the weather was not a big concern to us field geologists. Bad weather came with the job, and we just prepared for it by wearing the right clothing. However, I do remember that the coldest day I ever spent in the field was one winter day a few years later in southernmost Arizona, of all places.

Broken equipment, on the other hand, is something that did bother us, and Florida Mountain produced more than its share of malfunctions. In addition to the brakes failing on my pickup truck and the loss of its four-wheel-drive capability—no serious accidents resulted, thankfully—I had to contend with a broken push rod on the drill rig that shut us down for a number of days. On top of that, the drillers got stuck in the snow, twisted the drill rods off in the drill hole, and more than once lost fluid circulation in a drill hole, which prevented further drilling. Florida Mountain was a tough place to work. A state mine inspector even came by once and shut us down for some violation. I had to go to Boise to get that resolved. These events bothered me and were frustrating to this young, eager geologist because they delayed the purpose of the drilling in the first place—to see if ore existed below the top of Florida Mountain.

This project had other challenging moments as well. One evening George came back to the Idaho Hotel's dining room from the outhouse with a sheepish look on his face. When asked by his dining mates what the matter was, he said he had dropped his Brunton compass into the outhouse, and he had just spent a long time retrieving it. A Brunton compass is carried in a leather case attached to the geologist's belt and was used every day to accurately pinpoint one's location in the field and to measure the orientation of rocks. No one asked to borrow George's compass after that.

Then, in September, shortly after George's exploration of the bowels of the outhouse, I awoke with a toothache. It turned out my wisdom teeth were impacted. Oh, great! That was just what I didn't need. I drove into Nampa, Idaho, on the outskirts of Boise, located a dentist, and had two of them pulled on one side of my mouth. A month later, I returned and had the other two pulled.

During a November break from drilling at Florida Mountain, I flew once again on Midwest's Bullet from Boise back to Denver, then flew commercially to El Paso, Texas, and drove north in a rented vehicle to the town of Truth or Consequences, New Mexico. Here I joined two of Midwest's geologists whom you've already met, Dave Jonson and Dick Kehmeier. Since I was single and had no strong attachments to any one place, I was often called upon on short notice to take a trip somewhere to examine a prospect. In this case, the three of us examined a fluorspar occurrence in the Salado Mountains of central New Mexico, west of the Rio Grande River (I know that's redundant) and not far from the town of Truth or Consequences.

Fluorspar, the commercial name for the mineral fluorite (calcium fluoride or CaF_2), is used as a fluxing agent in the steel-making industry—it reduces the melting temperature of iron ore. We made a cursory evaluation of the property, collected samples for analysis, and ultimately determined that the prospect had merit and deserved more exploratory work. Consequently, Midwest acquired a lease on the property from a local prospector who had staked a number of mining claims there. This project would later become the subject of my master's thesis at the University of Western Ontario.

The noun "prospect" refers to a potential site of a mineral deposit or an area that has been explored in a preliminary fashion, often by digging. Prospects can sometimes be identified by the presence of an old mine shaft (a manmade, vertical passage dug from the ground surface for exploration or mining) or an adit (a horizontal excavation dug into a hillside) or just pits where old-timers dug in search of whatever. The Salado Mountains constituted a "raw" prospect because no digging had been done. However, Allen Sphar, the local prospector, had recognized fluorite outcrops over a wide area, and he knew the importance of them, so he staked some claims. An outcrop is rock that is exposed at the ground surface—that is, is not covered by soil.

After examining the Salado Mountains property, the three of us drove to Silver City (in southwestern New Mexico this time, not Idaho) and visited Kennecott Copper Company's Santa Rita copper mine. Also referred to as the Chino open pit, it was once the largest open-pit mine in the world. Looking into a manmade pit almost a mile wide was a thrilling experience. I had never seen a copper mine before, and this was an excellent example of a type of ore deposit that exploration geologists are trained to discover. Native Americans first "mined" this deposit as far back as AD 1100 when they collected turquoise and chrysocolla, both brightly colored copper-oxide minerals, for ceremonial decoration. The Spanish began mining copper veins there in 1799, and they could not have known how large the mining operation would become when full-scale mining commenced one hundred years later. Amazingly, the mine had been in continuous operation since 1910.

The drive to Silver City took us through the New Mexico Highlands, an extensive, mountainous area of fairly recent volcanism, only about thirty-five million years old. Hot ash had spewed out of volcanic vents and blanketed several hundred square miles of southwestern New Mexico. This six-hundred-foot-thick unit of ash became known as the Kneeling Nun Tuff, a volcanic deposit similar to that which destroyed the Roman city-state of Pompeii in AD 79. (Tuff is a rock composed of compacted and cemented volcanic ash.) The name Kneeling Nun Tuff comes from the topographic feature overlooking the Chino Pit that resembles a nun kneeling at an altar.

Kennecott Copper Company's Santa Rita copper mine (Chino Pit) in southwest New Mexico, 1972.

This was my first visit to the state of New Mexico, and during the next few years, I spent a lot of time in "The Land of Enchantment." I discovered that New Mexico is a land of contrasts: parched deserts and forested highlands, colorful flat plateaus (large, relatively high areas), sharp volcanic peaks, sand dunes, and the Rio Grande River. How different it was from New Hampshire! I loved the adventure of experiencing new places and deciphering how geology created the landscapes and ore deposits I was seeing.

Perhaps my love of adventure was sparked by reading *National Geographic* magazine in my youth where two of my favorite adventurers were often featured: Louis S. B. Leakey and Jacques-Yves Cousteau. Leakey intrigued me with his archaeological excavations of humanoid fossils in the Olduvai Gorge of Tanzania, Africa. *Maybe I'll go there someday*, I remember thinking. And Cousteau's exploits as the most famous undersea explorer of the twentieth century and inventor of the SCUBA device made me want to see more of the world. Interestingly enough, Cousteau's first experience at underwater diving was at Harvey's Lake in

Barnet, Vermont, only about ten miles from Bath, where I grew up! While attending summer camp there at the age of ten, he experimented with staying underwater by breathing through hollow reeds he found growing in the shallows of the lake. I've been to Harvey's Lake, but I didn't know of the Cousteau connection until many years later! Even though I had no childhood dreams of becoming a geologist—I didn't even know what one was—and I did not collect rocks as a youngster, I did dream of adventure.

In early November, I attended an annual meeting of the Geological Society of America in Minneapolis, Minnesota. The field of exploration geology is continually advancing, so geologists require ongoing educational updating; over the years, I attended as many such professional meetings as I could.

Upon returning from the meeting in Minneapolis, I traveled back to Florida Mountain to complete the drilling program before the major winter storms hit. Once done, I packed up the drilling samples in a U-Haul trailer and pulled it behind my pickup truck to Denver. On the way, I had a geologically and historically educational trip looking at the scenery through my truck window and reading about its geology at night. I had seen this territory from an airplane, but this was my first time traversing the Rockies on a highway, and it was a thrill.

I spent one night in Montpelier in the southeast corner of Idaho. Located in the Bear Valley, the town is on the route of the mid-1800s Emigrant Trail as it winds its way to Oregon and California from Missouri. This was probably one of the most beautiful places in which the emigrants stopped; at least it looked beautiful to me. I was surprised to learn that Brigham Young, who established this Mormon community, named it after the capital of his home state, Vermont.

From Montpelier, I drove east through mountains of the Overthrust Belt along the Idaho-Wyoming border. To understand the origin of these mountains, it helps to visualize the folds in a carpet that has been pushed against a wall. In this case, the Earth's crust was pushed from the west and bent into north-south-trending ridges. Some of the folds broke along broad, ramp-like faults, and slabs of rock thousands of feet thick slid east for as much as one hundred miles; they became stacked on top of one another like shingles. This is typical western geology spectacularly on display without the inconveniences of soil and vegetative cover that one experiences in New England. The Overthrust Belt's geologic terrain of stacked, faulted, and folded sedimentary rocks is a major oil- and gas-producing region; I saw many lonely oil-field pump jacks beside the highway, pumping away in a rhythm like that of the second hand on my watch but more slowly.

The subject of geology is a challenge to understand, not only because of the vast amount of time it took for our Earth to evolve but also because the spatial scales are orders of magnitude larger than those that people experience in everyday life. One must be a spatial as well as temporal thinker. It is almost impossible for untrained people to comprehend geologic concepts that involve structures tens of miles long—such as these overthrust faults—and events such as tens of thousands of fault movements of only fractions of an inch at a time. I remember a geology professor once telling his students that a particular mountain-size mass of rocks in Vermont had been transported nearly fifty miles to the east along a fault. I thought at the time, *You've got to be kidding! How can that happen?* Geology is nonintuitive. It requires solving puzzles, and that is one of the things that appealed to me most about the subject.

As I drove on and continued to marvel at the geology exposed outside my truck window, I could see Wyoming's Wind River Range off in the far distance to the north. This mountain range has the distinction of having the largest amount of displacement (movement) on its bordering fault of any range in the Rocky Mountains—44,000 feet. What this means is that the same rocks exposed on the nearly 14,000-foot-high peaks can be found on the other side of the fault thirty thousand feet beneath the adjacent Green River Basin. Imagine having nearly *eight miles* of fault displacement in your backyard. In this case, it took thirty million years and many thousands of earthquakes to accumulate the uplift of the Wind River Range.

My last stop before Denver was in Rawlins, America—as my two brothers-in-law who live in Rawlins, Wyoming, call it. Nearly everyone who travels across southern Wyoming has to stop in Rawlins to buy gas. In Denver, I delivered my precious samples to Skyline Labs, Inc. for analysis.

I didn't stay in Denver for long, however. Although winter weather prevented drilling in Idaho, we could drill in New Mexico. So in early December, Geoff sent me to Truth or Consequences to start drilling the Salado Mountains fluorspar property. There I worked with prospector Allen Sphar who had staked the first mining claims and from whom Midwest had leased the property. Allen was originally a uranium prospector and miner during the 1950s and 1960s who had mined in the Grants, New Mexico, uranium district. He told of exhaling into a Geiger counter just to hear it click. He also told of being lowered into one-thousand-foot-deep, five-foot-diameter drill holes at the Atomic Energy Commission's (now the US Department of Energy) Nevada Test Site to excavate caverns at the bottom for nuclear bomb emplacement. Allen was an amiable older guy, and I loved hearing him tell his mining stories.

Another colorful guy was Ralph Forsythe, who worked as a sampler for us on the drill rig. He had retired to Nogal, New Mexico, after a career running a circus, and he collected drill samples for us just for fun and to keep himself busy. Although stooped over a little, he thoroughly enjoyed manhandling the heavy samples in New Mexico's great outdoors. He and Allen were real characters, and I enjoyed working with both of them. Allen even invited me to watch the Super Bowl with him and his family one year.

Just before Christmas, I flew to back to Denver, then to London, Ontario, to visit the University of Western Ontario for the first time. Geoff Snow and Dave Jonson of Midwest told me how impressed they were with the consulting services Professor Bob Hodder provided to them, and it was their opinion that the University of Western Ontario was one of the best schools for exploration geology in North America. They knew that an additional college degree would help advance my career, and they suggested I take a look at Western Ontario. While in London, I met with Bob Hodder and other professors and with graduate students in the geology program. I was impressed with everything I saw and learned, and I decided that I would apply for admittance to their graduate program.

From London, I went to New Hampshire for the Christmas holidays and was back in Denver by the end of the month. Another fascinating and exciting year had passed quickly. What would the next year have in store for me?

4. I'M CANADIAN, EH?

At the start of 1973, I returned to Truth or Consequences to conduct more exploration at the Salado Mountains project. As project geologist, I supervised drilling (that is, told the drillers where to drill and how deep), geologically logged drill core, mapped the surface geology, prospected, and staked additional mining claims. In effect, I was in charge. All of this was in an effort to discover the extent of fluorite and determine its concentration.

My home-away-from-home from January through July that year was the Elephant Butte Inn in T or C, which is what New Mexicans call the town of Truth or Consequences. The Elephant Butte Inn is an appropriate place for a geologist to stay because the topographic feature after which it is named is a volcanic neck or plug that overlooks the nearby Elephant Butte Reservoir. The staff came to know me pretty well at the Elephant Butte Inn. In fact, when I checked out for the last time in July, the staff at the motel treated me to dinner and drinks. What a nice gesture!

For seven months my routine consisted of working in the field during the day, returning to the motel at night to clean up, getting something to eat—usually at the motel restaurant—and then returning to my room to compile data or review geologic literature. This schedule was the same on Saturdays and Sundays, and I didn't get out much otherwise. Although not antisocial, I usually didn't have an opportunity to interact with many people other than those working with me on the Salado Mountains project. Of course, I made conversation while ordering breakfast and dinner at a restaurant or getting gas for my truck, but that was about it. Even though I was alone a lot, I didn't mind it. I was comfortable with my own company, and I didn't suffer from loneliness. I found joy in being alone with the geology. Being able to survive the isolation is vital for an exploration geologist, and I learned that this kind of life suited me. After I was married, however, I did find it difficult being away from my family for extended periods. This was a contributing factor in my decision years later to discontinue doing exploration work.

The town of Truth or Consequences, a comfortable place from which to work, got its name from a popular NBC radio and television program of the same name. In 1950, Ralph Edwards, the host of the show, announced that he would air the program from the first town that renamed itself after his quiz show. And yes, the citizens of Hot Springs, New Mexico, took him up on it. Ralph Edwards came to town during the first weekend of May for the next fifty years to host an event called "Fiesta," complete with a parade, rodeo, and beauty contest. I even attended the festivities once. The Fiesta is still celebrated every year according to the T or C web page. Over the years since 1950, numerous attempts have been made to change the town name back to Hot Springs, but all have been unsuccessful. I think that's good.

With a population of around five thousand, T or C is the largest town in, and the county seat of, Sierra County. The town lies within the Rio Grande Rift, which extends the full length of the state from north to south and literally splits the state in half. This major break in the Earth's crust runs from southern Colorado through New Mexico to Texas, ending up in the Mexican state of Chihuahua. The Earth's crust in this rift is being pulled apart horizontally and is slowly dropping down between irregular fault zones on each side.

Standing there in T or C and looking around at the sagebrush covered ground, I could visualize the thousands and thousands of feet of sand and gravel that filled that rift beneath my feet. Geology is deceptive here, however. The Rio Grande River valley is much deeper than it appears to the casual observer. In places, the Earth's crust has dropped twenty-six thousand feet (about five miles) since rifting began here some thirty million years ago. There has been plenty of time for alluvial deposits to wash into and fill the rift. The Rio Grande River flows slowly within the rift from its source in Colorado's mountains; fields of cotton, nut trees, and chili peppers dominate the scenery along the river. This is North America's equivalent of the geologically more famous East Africa Rift Zone. Rifts as large as the Rio Grande are relatively rare in the world; I felt privileged to be able to study this one.

I've heard that Geronimo, the great Apache chief, often took his warriors to the hot springs at T or C to wash and heal their wounds. There is even a Geronimo Springs Museum in town. This made me realize that it wasn't so long ago that American Indians freely roamed this New Mexico countryside that I was now exploring.

One day in February, Midwest geologist Steve Zahony, consulting geologist Al Perry, and I drove to the Winkler Anticline in the Animas Hills in the southwest corner of New Mexico. This part of the state is called the "boot heel" because that's what its shape resembles on a map. Our purpose was to look at another fluorspar prospect that had come to our attention. We found fluorspar filling open spaces in black jasperoid (silicified limestone), and in some places masses of it were up to a foot across. This discovery was exciting, but what I remember most clearly was that the weather was terribly cold and rainy that day. So in midafternoon, we stopped at a saloon in lonely Hatchita, New Mexico, and bellied up to the bar for a shot of whiskey. We soon felt much better. This was just another first experience for this young geologist.

Here in the Animas Hills, I learned an American history lesson. We were close to the Mexican border where about six hundred Mexican revolutionaries under the leadership of General Francisco Pancho Villa surprised and raided the town of Columbus, New Mexico, in the dead of night on March 10, 1916. This was during the Mexican Revolution, and until the World Trade Center attack of September 11, 2001, the raid on Columbus was the most recent foreign attack on mainland US soil. Interestingly enough, Pancho Villa's youngest son died in 2010 in Hayward, California, not far from where I live in Dublin.

I was able to take a few days off from work in the Salado Mountains to experience a different kind of geology when I accompanied a field trip conducted by the University of Western Ontario's geology department for their students to learn by seeing ore deposits in the southwestern United States. I was pleased to have been asked to join them; more lifelong learning was underway for me. My participation also benefitted the students, since they could learn through conversations with me what it was like to be a practicing exploration geologist.

Geoff Snow was also invited, so he and I flew to Tucson and joined the students on the bus that took us to the Silver Bell, Sierrita, Esperanza, Twin Buttes, Pima, Mission, Ray, Morenci, and Metcalf open-pit porphyry copper mines of southern Arizona. (The term "porphyry copper" refers to the fact that the copper minerals commonly occur in intrusive rocks that are porphyritic—that is, they contain some large mineral crystals within a matrix of smaller crystals.) I quickly realized what a fabulous opportunity this was for me to study such a large collection of copper ore deposits. Reading about the deposits in a textbook was fine, but there was nothing better than going into a mine and putting your face to the rock exposures to experience what they had to teach.

Porphyry copper deposits contain low concentrations of disseminated and vein copper, usually less than 0.5 percent. That means every ton of rock mined contains, on average, only about ten pounds of copper. These mines are profitable only because of the large volume of ore that is mined by highly mechanized means. The Morenci mine was especially fascinating, not only because a huge mineral deposit is being mined there but also because of its thick blanket of enriched copper ore found near the ground surface. This zone is produced by supergene enrichment, which means chemical action by surface water substantially increased the copper content. The ore was so valuable that Phelps Dodge, the company mining the deposit, dismantled and moved the town of Morenci to reach the ore lying beneath the town site.

This cavalcade of young geologists also went to Jerome, Arizona, to examine the massive sulfide deposits there and at nearby Iron King. (Massive sulfide deposits are layers of almost pure metallic sulfide minerals within a sequence of volcanic rocks.)

As I learned that the Spaniards found outcrops of copper here in the 1500s and the American Indians used the blue and green copper oxide to make face paint, I was reminded that geology has been important to mankind for many, many years. You could consider the American Indians to be some of the first "geologists" to find and use this natural resource at Jerome.

Arizona is the largest copper-producing state in the country, so it is the obvious place for geology students to see what copper orebodies look like. Except for copper production at the Butte open-pit mine in Montana, at the huge Bingham Canyon mine outside of Salt Lake City, and at a couple of mines in Nevada, all US copper production at that time came from southern Arizona.

I enjoyed the enthusiasm and excitement of the Canadian students, as most had never been to Arizona before. What a great geologic and cultural treat for them, as it was for me. I enjoyed their company immensely.

After a wonderful week of learning and teaching, I returned to the Salado Mountains to continue drilling. One of the first things I did was to begin growing the full beard I still have today. My wife of thirty-eight years has never seen me without it. I grew the beard because I looked young for my age, and I thought I didn't command enough respect from the drillers. I think the beard did help because the drillers seemed to have more confidence in my decisions after that. Perhaps I just felt more mature and confident with the beard and therefore came across that way.

I worked out in the open at our drill rigs during an unusually cold and wet period that winter in the Salado Mountains. At one point, six inches of snow covered everything, and both the drill rig and my truck became stuck. What else could I do but dig my truck out of the snow and go skiing at the Sierra Blanca ski area on the Mescalero Apache Indian Reservation near Ruidoso, New Mexico?

En route, I discovered the White Sands National Monument in the Jornada del Muerto (Journey of Death) desert east of Las Cruces. White Sands is a collection of brilliant white sand dunes composed of glistening gypsum crystals, not quartz, as is usually the case. This is the world's largest gypsum field, and I had never heard of it. What a geologic surprise this skiing trip was turning into. The source of the gypsum is in the nearby San Andres Mountains where 250-million-year-old sedimentary rocks containing gypsum are dissolved by rainwater. With its dissolved load, surface water flows into a nearby basin where the gypsum is, in turn, precipitated in a mostly dry lake bed. When the wind blows, it picks up the sugar-fine gypsum crystals and heaps them into dunes up to fifty feet high. I hiked to the top of a dune and was intrigued to see a dramatic color contrast where the bright white dunes lie adjacent to black basalt of the nearby Malpais Lava Flow—stupendous beauty provided by Mother Nature's geology. This is certainly one of the world's most unusual natural wonders.

White Sands is also the site of the world's first atomic-bomb test, called Trinity, conducted on July 16, 1945. Led by Robert Oppenheimer, this test at the White Sands Missile Range was a key part of the US Army's top-secret World War II Manhattan Project. I can only wonder what the local ranchers thought early that morning when they saw a bright flash in the sky followed by a giant mushroom-shaped cloud. As large as this explosion was, however, it was a relatively low-energy event compared with the energy produced by earthquakes felt along the Rio Grande Rift. I didn't go onto the test site because it is open to visitors only two days a year. By the way, the skiing was excellent once I got to Sierra Blanca. What a pleasant and educational snow day for me!

At one point later in the year, Steve Zahony and I flew low over the Salado Mountains area in a Piper Cherokee airplane looking for more fluorspar prospects. I know most people do not like flying in small planes, but I did. You can see so much more than from a commercial airliner. Exploration from small airplanes turned out to be an unexpected and fun part of being a geologist. A couple of years later, I took flying lessons in Denver for a while, but I gave it up because I realized flying would be an expensive hobby.

Further adding to my experience database, I took one day to visit an oil-drilling rig in the Jornada del Muerto; this was the first oil-drilling rig I had ever seen, and it was huge. I spent part of the day on the deck of the rig talking to the drillers and roughnecks and learning about oil drilling. I recognized wealth being created when I saw it, and geology was the key.

Later on I attended a meeting of the American Institute of Mining, Metallurgical and Petroleum Engineers in Tucson and toured Duval Mining Company's Sierrita open-pit copper mine south of town. I always went on any mine tours that were available. Besides the geologic exposure that I received on these tours, I was always fascinated by the huge electric shovels that dug the ore and the diesel-electric haulage trucks that carried the ore out of the pits. I suppose this fascination was the little boy in me coming out. I wished I could drive one of those huge trucks, but of course I never did.

The hubs of the wheels on some of the trucks were as high as I am tall. A diesel engine in each truck powers a generator that produces electricity, which is then fed into an electric motor in each wheel. These motors turn the wheels. Because this equipment is so big, the haulage truck driver's visibility is severely limited; thus, smaller vehicles like pickup trucks sport long antennas with bright red flags. Without them, the pickups ran the risk of being run over by a truck carrying two hundred tons of ore. Such an incident did happen once, and the driver of the ore-haulage truck didn't even feel the impact. I don't recall hearing about the fate of the pickup truck driver, but I don't see how he could have survived.

During the early part of 1973, I took a few days off from the Salado Mountains and flew back and forth from Albuquerque to Denver several times. This was so I could go cross-country skiing and downhill skiing with friends at Colorado's Loveland, Arapahoe Basin, and Keystone ski areas. Jeff Stimson, a friend I grew up with in Bath, New Hampshire, had recently moved to Denver. The two of us had a particularly exhilarating time skiing off the cornices at Arapahoe Basin one sparkling Sunday, and the following day we arrived at our respective places of work sporting brightly sunburned faces. Nevertheless, we both agreed the exhilaration of that day's skiing was worth the subsequent discomfort! Since I was alone so much on the job, it was important for me to do some socializing when I was back in Denver, and I thoroughly enjoyed the camaraderie I developed with my skiing partners.

On one of those trips back to Denver, I toured the underground Climax molybdenum mine near Leadville, Colorado. This was the largest molybdenum mine in the world at the time, and from 1925 to the 1970s, it produced more than half of the world's molybdenum. This was an important mine for the country's World War II effort because it had supplied virtually all the molybdenum that toughened the steel of the Allies' armor and weaponry. Three orebodies overlap vertically at Climax, each in the shape of an inverted cup lying above three separate intrusions (masses of igneous rock that intruded preexisting rock). Amazingly enough, a major fault cuts one orebody, called Ceresco Ridge, producing a deep offset portion more than ten thousand feet below the ground surface.

I cherish the beautiful specimens of silvery-gray molybdenite (MoS_2) ore that I collected at the site. Every geologist I know has a precious (to him/her) collection of rocks and minerals that he/she takes with him/her every time he/she moves and would not part with for anything. Specimens in geologists' collections are often rare, unique, or particularly good examples of rocks or minerals, and they represent the effort a geologist has expended to collect them. One could buy such samples at a gem and mineral show, but a geologist would rarely do so.

The Climax mine is located high on the western slope of Colorado's Tenmile Range, and a geologist at the mine told me that because of the high elevation and low air pressure, water boils at a lower temperature there than at sea level. Consequently, he could safely drink his morning coffee while it was boiling, he said. Whether he was pulling my leg or not, I do not know. But it sounded reasonable. Not being a coffee drinker at the time, I chose not to try this experiment; but being a scientist, I should have. The community of Climax, which is where many of the miners lived, is the highest-elevation settlement in the United States at 11,360 feet.

In early April, we completed rotary and core drilling at the Salado Mountains, and everyone left except me. In the meantime, I had been accepted into the University of Western Ontario's geology graduate school and planned to begin my studies there in September. With Professor Hodder, I had been discussing by telephone and mail various possibilities for a master's thesis research project. Unraveling the genesis of the fluorite in the Salado Mountains was one possibility. So when drilling finished, Bob Hodder flew from London, Ontario, to El Paso, Texas, where I picked him up at the airport and took him to T or C. Hodder spent a couple of days examining the geology in the field with me, and we talked late into the night about what we did and did not know about the geology.

At the end of those two days, he heartily agreed that there were enough mysteries about the fluorite in the Salado Mountains that geologic mapping of the entire mountain range would be an excellent research topic for a master's thesis. Detailed mapping would hopefully allow me to determine the genesis of the fluorite. Furthermore, Hodder agreed that I could do the mapping that summer *before* even enrolling at the university in the fall. Bob Hodder is like Geoff Snow in that he willingly and eagerly shares with others his geologic expertise, and he quickly became another mentor of mine.

I was now alone on the project, and I immediately began detailed geologic mapping of six square miles of the Salado Mountains, continuing into early July. Surprisingly, I was initially a little lonely since my coworkers had left, but once I got into the swing of mapping, the anticipation of what I would see next kept me fully occupied. I seemed to be coaxed along every day by a need to know what was beyond the next bend. I had no distractions and could fully concentrate on my thesis work since Midwest Oil continued to pay me for the rest of the summer; they hoped that my work would lead to additional discoveries of fluorspar.

Geologic mapping consists of plotting on a topographic map or aerial photograph the locations of rock types and geologic

Young geologist Lamarre in the Salado
Mountains of New Mexico, 1973.

structural features—like faults and folds—that crop out (are exposed) at the ground surface. On my thesis map, I portrayed each rock type in a different color or pattern. Where there were no rocks exposed, I extrapolated between nearby rock exposures. The purpose of mapping is to visualize in three dimensions what the subsurface geology looks like. Being capable of three-dimensional-spatial thinking is a vital skill for a geologist, and I was fortunate to be able to do it; not everyone can. Without this ability, a geologist would be severely handicapped. For instance, if you were to find a quartz vein on the surface, you would want to know if it continues vertically down into the subsurface or if it angles off in some other direction. By plotting where the vein crops out on the surface, you would be able to make that determination. In essence, after some study a geologist must be able to look at a hillside or mountain range and visualize what is happening below ground. Because this is difficult to do, field geologists sometimes need the patience of a rock to decipher the clues.

From April to July, I parked my truck in the morning and spent the day scouring the ocotillo- and prickly-pear-covered hills as the rocks in the Salado Mountains began to speak to me. Fluorite there occurs as fracture fillings in jasperoid, which is silicified carbonate rock. The jasperoid is distinctive because of its reddish-brown color and the knobby outcrops it forms due to its resistance to erosion; therefore, it was relatively easy to map.

I was pleased to be the first person to identify and map Precambrian rocks in this mountain range, although they had nothing to do with the genesis of the fluorite. At 1.6 billion years old, these rocks are some of the oldest in New Mexico, yet they are less than one-third the age of the Earth. In the overlying, younger sedimentary rocks, I was especially excited to find well-preserved fossils—trilobites, crinoids, and brachiopods. As I sat and stared at the fossil-containing rocks, I tried to imagine what the landscape looked like during the Paleozoic era (251–542 million years ago) and how these now-extinct creatures fit into it. They got to see the terrain so much earlier than I did. By the way, it was while contemplating the geology in this one small part of the world that I developed my fondness for Spam, which I commonly carried for lunch. Combined with crackers and an apple, a small can a day is all you need. And you can buy it in bulk.

In hindsight, it was probably not prudent for me to be out there mapping by myself. Had I been injured, it would have been days before anyone realized I was missing. As I remember, the property was about a one-half-hour drive west from town, and no handy cell phones existed then. I didn't give it much thought at the time though; I just went out to the property every day and went to work. However, I did have a close call one hot day when I became weak and was barely able to walk back to my pickup truck. My legs just gave out on me. It was one hundred degrees, my first experience with triple-digit temperatures, and apparently I had not been drinking enough water. Heat exhaustion was the result. My wife says I did not learn a lesson that day because I still do not drink as much water as I should. A few years later, I realized that this particular day was a relatively mild one compared to working in 110–115 degree summer heat in the Sonoran Desert of southern Arizona.

I had one near disaster of a different sort one day when the ignition switch in my truck shorted out as I was driving through town. Flames burst out under the dashboard, and smoke filled the cab. I immediately stopped the truck on the side of the highway and, gathering my thesis map from the seat beside me, placed it carefully on the ground beside the truck for safekeeping. Having it burn up in a truck fire was definitely not an option. After extinguishing the small flame and clearing out the smoke, I drove off. Shortly thereafter, I looked at the seat beside me and, to my horror, saw that my geologic map was missing. This map was the result of four months of hard work; I could not lose it. I quickly turned the truck around in the middle of the street and hastily returned to the site of the fire. Thank heavens, the map was still on the ground where I had carefully placed it.

I had another close call one morning when I arrived at my destination ready to begin fieldwork. As I opened the door of my truck and swung my legs off the seat, my mind was focused on the day's mapping ahead and on nothing else. Consequently, I almost stepped on a rattlesnake coiled up right below the door. After that incident, I never again got out of the truck without first looking at the ground before putting my feet down. And while working in the field, I always looked for potential hideouts where I knew rattlesnakes might be. So far, no mishaps!

In mid-June, I took a day off to fly to San Francisco to consult with Canadian immigration officials regarding my move to Canada in September to attend graduate school. With my nose plastered to the airplane's window on the way there, I had spectacular views of the Grand Canyon, one of nature's geologic wonders. Seen from an airplane, one can appreciate in the canyon's length and depth the immense erosive power of the Colorado River. Viewing this marvel was another first for me. I returned to T or C having successfully completed the Canadian immigration paperwork.

Toward the end of my mapping, I invited Dr. Noel McAnulty, Sr., a professor in the geology department at the University of Texas at El Paso, to take a look at my thesis area with me. Having a longstanding interest in fluorspar, he eagerly accepted my invitation. Dr. McAnulty had worked as an exploration geologist for Dow Chemical Company and was a recognized expert in fluorspar exploration. I was honored that he agreed to spend time with me, and I was beginning to feel that I could make a contribution to the science of geology. I had my own thoughts, of course, but Dr. McAnulty gave me some suggestions about what he thought the geologic conditions may have been that caused fluorspar to be present in the Salado Mountains. I gave careful consideration to his suggestions over the next year and incorporated some of them into my thesis. Finally, on July 7, 1973, after completing mapping and tying up loose ends in T or C, I packed my belongings into my pickup truck, said good-bye to the Sphars, and started driving north to Denver, having completed my thesis fieldwork.

In Midwest's office, I began compiling my thesis data, drafted maps, and organized my thoughts about Salado Mountains' geology. Then on August 15, I started driving home to Bath, New Hampshire, to visit my family. This drive across the country was along Interstate 80, a route I traveled many times over the years. I then drove to London, Ontario, where, on September 2, I became an expatriate in Canada and took up residence there. I was ready to begin geology graduate school at the University of Western Ontario.

Returning to the classroom was a little strange at first, but I got over it quickly as I learned that all my professors were experts in their fields and good instructors. I also discovered that my fellow graduate students were as eager to learn as I was, so we got along well.

One of the first things I had to learn was the Canadian speech trait of putting an "eh" at the end of their sentences. Secondly, any geologist worth his salt knows the geology of the place where he lives, so I had to learn about the geology underlying the campus. London, Ontario, is about halfway between Detroit and Toronto, and it lies on flat Paleozoic sedimentary rocks of the stable interior platform of the North American continent. These limestones, sandstones, and shales extend from the Appalachians on the east to the Rockies on the west. These sedimentary rocks are 3,500–20,000 feet thick and were deposited on top of older Precambrian basement rocks in shallow and warm tropical seas. This geologic environment of marine and river deposits has been quiet and stable since sedimentation began more than 650 million years ago, hence the term stable interior platform. Maybe this is why Midwesterners are such stable citizens.

Because there are no hills or mountains around London, not much geology is visible. However, geologists have deciphered the geology by studying the rocks retrieved when oil and gas wells are drilled, of which there are many. In fact, it is only about 150 miles across Lake Erie to Titusville, Pennsylvania, where in 1859 Drake's Well was drilled, North America's first productive oil well.

Wasting no time, in late September, Professor Bob Hodder took a group of us economic geology students, both graduate and undergraduate, on a field trip to the geologically famous mining camps of Cobalt, Ontario; Noranda, Quebec; and Kirkland Lake, Ontario, about 350 miles north of London. The ore deposits there are in the Canadian Shield, a generally flat and seemingly endless expanse of eastern Canada covered with evergreen and white birch trees and swamp, and inhabited mostly by mosquitoes and black flies. Canadian geologists call this "the bush." Because the rocks are very old (2.5–3.8 billion years), they are referred to as "basement rocks," ancient crystalline and highly metamorphosed rocks that make up the foundation of all the continents on Earth. Identifying the original rock types that have been metamorphosed (changed by heat and pressure) into other rock types with funny-sounding names like schist, phyllite, and gneiss (pronounced "nice") is often difficult. In addition to forming the Canadian Shield, these old Precambrian rocks comprise the heart of many mountain ranges of the Rocky Mountains in Colorado, Wyoming, and Montana.

Cobalt, Ontario, our first stop, began as a silver mining camp when rich silver veins were discovered in 1903 during construction of the trans-Canada railroad. This historic mining town of about 1,200 people is the birthplace of hard-rock mining in Canada. By the end of the depression in the 1930s, the high-grade silver ore had been mined out, and Cobalt was well on its way to becoming a ghost town. However, someone realized that the metal cobalt, which was contained in the silver ore but had been a troublesome nuisance when smelting and refining the silver, was now in demand because of the threat of war and new technological needs. Thus, cobalt mining took the place of silver mining and kept the town of Cobalt alive.

In nearby Quebec, the Noranda mining camp began with the discovery of copper in 1917, and it grew to become the copper capital of Canada. The word Noranda is short for "Northern Canada" and is the namesake for what was the international mining company Noranda Mines, Ltd., a subsidiary of which I went to work for after graduating from the University of Western Ontario. Beneath the town of Noranda, copper, lead, and zinc are mined from bodies of massive ore called massive sulfide deposits. Imagine it: these deposits are 2.5–3.8 billion years old and are the sites of ancient under-sea volcanic activity. The Horne mine and smelter were Noranda's primary operations there, and all of us students were impressed with the variety of minerals in the Horne orebody. We collected many specimens to add to our personal collections; of course, that meant we had to haul them out of the mine on our backs.

Kirkland Lake, Ontario, is a famous gold-mining town along the Ontario-Quebec border, but it is probably more famous for the large number of National Hockey League players it produces. While looking at outcrops in the area, Bob Hodder was quick to point out the surface-water divide where water no longer flows south to the Great Lakes or the St. Lawrence River but instead flows north to Hudson Bay and eventually to the Arctic Ocean. We were, indeed, way up north.

Upon returning to London, I marveled that in just a few short days I had seen excellent examples of silver, cobalt, copper, and gold ore in their classic Precambrian geologic settings. *Where else could I have done that?* I wondered.

In early October, northern Ontario was the destination for another field trip, this time to the Sudbury nickel mines and Elliot

Lake uranium mines. Sudbury is a major Canadian nickel-mining center, and during World War II, one mine alone provided 40 percent of the nickel used for Allied artillery. Ore was discovered in Sudbury in 1883 during construction of the Canadian Pacific Railroad when construction workers inadvertently excavated into the orebody. Even though they were not exploration geologists, we can thank the railroad builders for many mineral discoveries.

The Sudbury ore deposits are part of a geologic structure known as the Sudbury Basin, believed to be the remnants of a 1.85-billion-year-old meteorite impact crater. Because impact craters are rare, I was thrilled to be able to collect samples of rocks exhibiting shock-metamorphic effects called shatter cones. They're also called spinifex textures because they resemble an indigenous grass by that name that grows in Australia. The cone-shaped, radiating fractures of shatter cones are produced in the rocks in response to the tremendous impact of the meteorite. You could call them fossilized shockwaves. At thirty-six miles across, this meteorite impact crater is the second largest on Earth. Because of erosion and heavy vegetation, however, the crater is not obvious to anyone casually driving around.

Nickel ore at Sudbury contains an unusually large concentration of sulfur, and smelting operations over the years have released it into the air where it combined with water vapor to form sulfuric acid. As a consequence, the vegetation had been devastated by the effects of acid rain. When combined with extensive logging to provide fuel for early smelters, the ecology was nearly destroyed. I could see brown trees and barren hillsides comprising much of the scene. Sudbury was not a pretty city.

There was something else that was not pretty either. Someone yelled, "Look! What's that up on those rocks?" Then someone said, "He doesn't have any clothes on!" Because it was a fad at the time, Harlan Meade, a fellow graduate student from British Columbia, was seen streaking the outcrop at Sudbury. He did have his field boots on, though! Bob Hodder just shook his head in wonderment. We all enjoyed the camaraderie of Bob Hodder's field trips, and this was an example of the fun we had.

The community of Elliot Lake is a relative newcomer to the mining scene in that the Blind River uranium district was established there in 1955 when uranium was discovered in Archean (the oldest part of the Precambrian eon, 2.5–4.6 billion years old) metasedimentary rocks. When I visited Elliot Lake, it billed itself as the "Uranium Capital of the World." It's a good thing it had that going for it because it sits in a lonely, remote location near the north shore of Lake Huron. Low-grade uranium ore occurs in quartz-pebble conglomerates (paleo-placers), and for many decades after World War II, these deposits—along with similar quartz-pebble conglomerates in South Africa—were the major source of the world's uranium.

Soon after our return to London, four of us—Bob Hodder, fellow graduate students Harlan Meade and Dorothy Atkinson (from England), and I—flew to Bangor, Maine, to examine the massive sulfide deposits on the Atlantic coast at scenic Penobscot Bay. Hodder knew the area well because he had been a geologist at the Black Hawk mine there while working for Callahan Mining Company. Hodder was the graduate advisor for the three of us, and he knew we would benefit from studying these ore deposits and their geology firsthand.

The most impressive feature of these Maine ore deposits was my realization that they were associated with old seafloor rocks—although now on land—that contain layers of almost pure sulfide minerals. We went underground at the Black Hawk mine, and I could almost envision hot, metal-rich liquids and gasses bubbling up from an under-sea volcanic vent into ocean water that caused cooling and precipitation of the metals onto the ocean floor.

As you can see, Bob Hodder was a fanatic about field trips, and at one of our stops in Maine, it was so late in the day that we had to look at the outcrop by the headlights of the car. All of his field trips that fall were wonderful: three in three months. These trips were instructive as well as fun, and they added immensely to our education and thoroughly enriched our graduate school experience—just what education should be. Hodder was teaching us to listen to what the rocks have to tell about their history, and we listened eagerly. After all, these outcrops provided a window into a world of another time. Hodder encouraged us to observe, feel, and smell the rocks in their natural setting so that in our mind's eye we could imagine how the ore formed. We had to pick up on the clues the rocks were sending us. Listening skills were being taught, even though we may not have called them that at the time.

Harlan Meade, Dorothy Atkinson, and I were fortunate to have Bob Hodder as our thesis advisor; we all looked up to him. By taking us on these field trips, he demonstrated his commitment to giving us the best exploration-geology education possible; he was truly interested in our well-being and seeing us succeed. Hodder willingly shared his wealth of knowledge with us and was always friendly and pleasant in doing so. In fact, Harlan and I even surprised him that Halloween by showing up unannounced at his house for trick-or-treating. It's too bad I don't recall what costumes we wore. I suspect his wife and two young boys thought we were crazy. Hodder himself was amused and just shook his head.

5. DEGREE IN HAND

I spent most of 1974 busily taking classes and writing my thesis. In my small, shared office in the geology department building—I spent many long days and nights there—I poured over geology textbooks and organized my thoughts for making a coherent story about fluorite in the Salado Mountains. Other graduate students were taking similar courses and doing their own thesis research, and we often met in the hallways to compare notes. I spent very little time in my little apartment on Cherry Hill Circle; I was there only to eat and sleep. Crafting a thesis is difficult, and I remember receiving Bob Hodder's review comments on an early draft—the pages were literally covered with his red-ink edit marks and comments. Although difficult to accept at the time, I was not discouraged and recognized this was part of the learning process. Hodder's close attention to my writing stood me well in the corporate world later on as I became a much better writer as a result of his conscientious efforts.

Even though my life at the time was entirely devoted to taking geology courses and cranking out my thesis, I also absorbed much of the local culture during my graduate school days. What I remember and appreciate most about the university and the city of London, Ontario, was the international atmosphere prevalent throughout. I could hear any number of different languages being spoken in a single day. Northern New Hampshire had little, if any, cultural and ethnic diversity, so that part of my education had been sorely lacking. At Western Ontario, however, my fellow graduate students hailed from such diverse places as England, New Zealand, Australia, Italy, Ghana, Brazil, the United States, and from all over Canada. One day I asked fellow graduate student Yaw Nitiamoah-Adjaquah from Ghana where he learned such good English. I was embarrassed by my ignorance when he replied, "It's my native language."

Even though I spent a year in Canada, I never did go to a hockey game or watch a curling match, both Canadian national pastimes. How did my Canadian friends let me get away with that? We all must have been too busy studying.

Since it was on the way, on my drive from London to New Hampshire to visit my folks for Christmas, I stopped at Niagara Falls to see that natural wonder. I learned that these huge waterfalls retreat about three feet each year as they erode their way upstream through the Niagara Escarpment, a cliff formed of resistant sedimentary rocks. The falls were born about 12,300 years ago as the continental ice sheet retreated. Since then, they have eroded their way eight miles upstream, creating an eight-mile-long gorge. I was witnessing geology in action. All the water in four of the Great Lakes flows over Niagara Falls on its way to Lake Ontario and the St. Lawrence Seaway. There was geology to see in this part of the world after all.

In early January, we "ore deposits" students—undergraduate and graduate—took the biggest field trip the University of Western Ontario offered. We traveled to Arizona to learn more about the ore deposits there. We toured some of the same open-pit copper mines that I had seen the year before on a similar trip, and visited others as well: Bagdad, San Manuel, Bisbee, and Cananea in northern Mexico about thirty-five miles south of the border. Visiting Cananea was especially challenging since we were engulfed in a snowstorm. On the return to the United States, we were pleasantly surprised when Bob Hodder, our leader, had us stop for dinner in Tombstone, Arizona, site of the most famous western gunfight in history at the OK Corral. I guessed that Tombstone didn't look much different in 1973 than it did on October 26, 1881, when the Earp brothers and Doc Holliday became famous for drawing guns on the outlaws.

University of Western Ontario geology Professor Bob Hodder
and fellow graduate student Yaw Nitiamoah-Adjaquah
examine a pillow basalt outcrop in northern Ontario, 1973.

When back in London, it was time to hit the books again. Then, in midsummer, I accompanied Bob Hodder on a visit to Harlan Meade's PhD thesis area in the Canadian Rockies of northern British Columbia where we stayed as guests in Harlan's field tent-camp. Harlan was a short, blonde-haired geologist who hailed from southern British Columbia; his enthusiasm for all things geologic was impressive and contagious. He later became a highly successful mining executive in Vancouver. I had never been to western Canada before, and I discovered, this country was remote, as remote as the Alaska Range. As I lay in my sleeping bag looking out from the tent, I could see no evidence of human activity, just heavily forested hillsides. The closest "town" was the unincorporated settlement of Germansen Landing, just west of the Rocky Mountain Trench and east of Ketchikan, Alaska. To reach Harlan's camp, Bob Hodder and I flew into Edmonton, Alberta, rented a car and drove west, and then hiked in. Harlan enthusiastically shared with us the geology he was unraveling, and I learned about a new geologic setting—the Canadian Rockies.

Then it was back to thesis work again. After addressing reviewers' comments on my thesis and furiously rewriting drafts during the summer, I submitted the final product to my thesis advisory committee in early fall. As you can imagine, this was done with considerable anxiety on my part. Much to my relief, the committee approved my thesis, "Fluorite in Jasperoid of the Salado Mountains, Sierra County, New Mexico," and I was awarded a master's of science degree in geology in September. I had overcome my misstep at the University of Wyoming, and graduate school at the University of Western Ontario had been a wonderful experience. I knew I had received a top-quality education. Regarding the Salado Mountains fluorspar deposit itself, even though we had identified 2.35 million tons of rock containing 19.35 percent CaF_2, this was not enough to constitute ore. Midwest did not put the deposit into production, and eventually they dropped the mining lease.

During the summer, I had written letters to many mining companies seeking a full-time job as an exploration geologist once I graduated, and I had received some attractive offers. After carefully reviewing them, I decided that an offer from my former boss, Geoff Snow, to work in the Colorado Rockies sounded the most appealing. Not only was the scenic terrain of the Rockies a draw, but I also looked forward to working with Geoff again. As I have said, he is an excellent teacher who eagerly took time from his busy days to explain facets of geology to me.

On October 1, 1974, with newly minted graduate degree in hand, I started my job as a full-time, permanent exploration geologist with Norandex, Inc., which Geoff headed. Based in Denver, Norandex was the US exploration arm of Noranda Mines, Ltd., which at the time was the second largest Canadian mining company, producing copper, lead, zinc, nickel, and gold, primarily from eastern Canada. Not long after my arrival in

Denver, Norandex changed its name to Noranda Exploration, Inc., and I worked for Noranda for the next ten years.

I was both excited and a little apprehensive to be joining the corporate world as a full-time professional. This was a big step for me. I looked forward with eager anticipation to applying my many years of training to discovering that elusive ore deposit. When I arrived at Norandex, I found that many of the people I had worked with at Midwest Oil Corporation now worked there. They welcomed me with open arms, and it immediately became apparent that they considered me, a new project geologist, to be a full-fledged professional.

On the very first day of my new job, Steve Zahony, now my immediate supervisor, and I drove up into the Colorado Rockies, parked our vehicle, and hiked about five miles into Willis Gulch in the Sawatch Range. We were conducting reconnaissance for gold. That term refers to a general geologic examination of an area, usually conducted as a step preliminary to more detailed studies. As I stopped to catch my breath I had been living close to sea level for the past year, and the Sawatch Range tops out above fourteen thousand feet—I admired the high-altitude splendor of that beautiful snow-covered mountain range set against a perfectly blue sky. What a fun way to start a new job!

I especially liked working with Steve, also. A lanky swimmer from Ohio State University who received his master's degree in geology from Dartmouth, he talked often about his parents' difficult days in Hungary under Communist rule. This was at the height of the Cold War, and Steve's dislike of the Soviet Union was clear. I had been pretty oblivious to world politics, so listening to Steve's stories was a real eye-opener. I now had an appreciation of what life was like living in a Communist society, something I had no concept of before.

To reach the Sawatch Range, we drove up into the Rocky Mountains from Denver, which sits on the edge of the Great Plains at the Rocky Mountain Front. At that mountain front, the Rockies rise abruptly into the sky just west of Denver, forming a north-south barrier to travelers. Peak elevations in Colorado easily reach ten thousand feet, and the state boasts fifty-four summits higher than fourteen thousand feet. Fifteen of these summits are in the Sawatch Range, one of them being Mt. Elbert (14,433 feet), the highest peak in Colorado. We could see it from our perch at Willis Gulch.

High in the Sawatch Range, Steve and I staked a number of mining claims on some attractive-looking geology. The backbone of the mining industry is the federal Mining Law of 1872, which authorizes and governs prospecting for and mining of economic minerals on federal land. All citizens of the United States eighteen years old or older have the right to explore for, discover, and mine valuable mineral deposits on public domain. Virtually all of the exploration work I conducted over the years was done on public-domain land controlled by either the US Forest Service or the US Bureau of Land Management. Exploration on National Park Service land or in wilderness areas, however, is not permitted. We staked our claims at Willis Gulch on Gunnison National Forest land.

A mining claim encompasses a rectangle of land 1,500 feet by 600 feet whose boundaries are usually marked by four-inch by four-inch by five-foot-long wooden posts set at each corner. Sometimes blazed trees are used instead of posts. A location notice must be posted on each claim, giving the date of location of the claim, name and address of the locator, and name of the claim. Names of the claims are left up to the claim staker, and often they are tied to some nearby geographic feature and numbered sequentially, such as Gold Peak #1, Gold Peak #2, Gold Peak #3, and so on. This doesn't mean that an ore discovery has been made, just that a parcel of land has been reserved for exploration by the claim staker. No land ownership is conveyed to individuals or companies; the US government retains ownership, but companies or individuals may conduct mining there.

Staking claims is time consuming and hard work, and we sometimes contracted that out. There are companies that specialize in claim staking; most of their employees are strong, young males. To be valid, once a claim has been staked, a location notice must be officially recorded in the county clerk's office of the county in which the claim lies. To maintain a claim, each year the claim owner must perform some kind of exploratory work on each claim, called "assessment work," and evidence of that assessment work must also be recorded in the courthouse. Consequently, I made many visits to county courthouses over the years to file claims and evidence of assessment work or to see if land was open for staking.

After more field evaluation of the geology at our Willis Gulch claims, the assay results (chemical analyses performed by a contract analytical chemistry laboratory) from rock samples we had collected turned out to be disappointing. Gold values were much too low to be economically viable, and we did no more work there. So I was then off to another project, this time in Utah.

I was given lead responsibility for the Lisbon Valley copper project south of Moab, Utah—population about five thousand—on the Colorado Plateau. This project was a joint venture with Centennial Development Company of Salt Lake City; Bob Gilmore was their onsite representative. The property consisted of copper-oxide and -sulfide minerals hosted by (contained in) sedimentary rocks. Centennial had previously drilled and delineated 6.3 million tons of ore, and Noranda's objective was to add to these reserves by defining a total of twenty-five million tons of 0.6 percent copper. I was excited to be working on this project. Imagine me, the new guy, leading a large drilling program for a major company.

During my first visit to the property in mid-October, which was to learn the lay of the land, I immediately saw that this was going to be a scenic place to work. In the distance was Arches National Park just north of Moab, and I visited the park one day to see some of the more than two thousand sandstone arches preserved there. Later on, I went to Canyonlands National Park southwest of Moab to see the pinnacles, balanced rocks, and fins that stun the senses of visitors, park rangers, and geologists alike. I was awestruck by these stupendous examples of ancient geologic miracle work. The town of Moab sits adjacent to the Colorado River, and a predecessor geologist, John Wesley Powell, and his crew floated past here in 1869 on their exploration of the Colorado River and the Grand Canyon. This is viewed by many as the greatest and best known exploratory survey of the American West.

From the Lisbon Valley property itself, I could admire the nearby stand-alone La Sal Mountains, rising 12,721 feet in the air. They seem to rise right out of the surrounding flat terrain of the Colorado Plateau, which is at an elevation of about four thousand feet. Located about fifteen miles east of Moab, the La Sal Mountains often form the backdrop for beautiful photographs taken from Arches National Park. The abundance of surrounding beauty easily distracted me while working; I often found myself gazing off into the distance, amazed at what geology had produced.

In addition to their natural beauty, the La Sal Mountains are of geologic interest because they are formed by an unusual type of geologic intrusion called a laccolith. When magma is injected between layers of sedimentary rocks deep below the Earth's surface, it separates the layers. If the force of the magma is strong enough, it pushes the overlying strata (layers) upward; that is, it causes the rocks to bulge and take on a domal or blister shape with a flat bottom. After millions of years of uplift and erosion, the laccolith may then be exposed to human view as a mountain.

Sadly, this intrusion is not a cactolith. Then I could give you this definition: "A quasi-horizontal chonolith composed of anastomosing ductoliths whose distal ends curl like a harpolith, thin like a sphenolith, or bulge discordantly like an akmolith or ethmolith." I'm not making this up! This is the definition from the American Geological Institute's 1972 edition of *Glossary of Geology*.[1] The term and its definition were coined by a US Geological Survey geologist to describe the nearby Henry Mountains. The definition was also meant as a tongue-in-cheek commentary on the absurd number of "-lith" words in the lexicon of geology. Who says science can't be fun!

---

1. Margaret Gary, Robert McAfee, Jr., and Carol L. Wolf, Eds., *Glossary of Geology* (Washington, DC: American Geological Institute, 1972).

Flying into Moab from Grand Junction, Colorado, on a small, propeller-driven commercial plane was always fun, especially one day when the pilot let me sit in the copilot's seat. Can you imagine anyone permitting that today! I didn't touch anything, but I was engrossed with watching the pilot control the airplane. On one flight back to Denver from Grand Junction in a larger Frontier Airlines jet, the pilot flew a complete circle above the Black Canyon of the Gunnison River so that we passengers could have a better view. We all immediately moved to the side of the plane tipped toward the river and marveled at the depth and length of the canyon. Although not as large as the Grand Canyon, it is still impressive.

Later in the fall, I went to Timmins, Ontario, for a Noranda company meeting and mine tour. This gold-mining town is even farther north in Ontario than Kirkland Lake, which I had visited during my graduate school days. Timmins is the hometown of country-rock singer Shania Twain, a favorite of mine in the 1990s, although she was not widely known in the 1970s when I visited her hometown. Listening to Shania brought out the sense of longing and beauty that I had experienced in my exploration work years before.

In nearby Quebec, we meeting attendees toured Noranda's Langmuir copper-nickel mine, the Pamour-Porcupine copper-gold mine, the Kerr-Addison gold mine, the Mattagami mine, and the Orchan mine. Orchan mined especially high-grade ore containing zinc, copper, silver, and gold. Thus, I had more exposure to rocks that contained orebodies.

Upon returning to Colorado from this trip to northern Canada, I began work on my second major project, investigating the Gem Park carbonatite (an igneous rock consisting primarily of carbonate minerals) in the Wet Valley near Westcliffe in southern Colorado. This prospect was brought to Noranda by geologist Dolf Fieldman of Congdon and Carey, Ltd. My future wife, Janet, was employed by Congdon and Carey at that time, although I had not yet met her. Congdon and Carey had conducted limited drilling at Gem Park and wanted Noranda's assistance in conducting additional exploration.

As project geologist, I began geologically mapping the carbonatite dike swarm that contains the important metallic element we were seeking—niobium. Niobium is an alloy metal that when added to iron in small amounts results in steel having significantly increased strength. This type of alloy steel is commonly used in jet engines. Because the rock exposure is so poor at Gem Park, I hired a local bulldozer operator to excavate a series of trenches to expose the rocks hidden below the soil cover, like we had done on the Florida Mountain project. I then began geologic mapping of the trenches.

While working there, the friendly, little ranching town of Westcliffe, population around four hundred and fifty, was my home. The town is in a nice setting with the jagged, snow-crested ridge of the Sangre de Cristo Mountains rising in the west and the lower Wet Mountains to the east. Of course I didn't know a soul in town, but that didn't matter. I stayed in a small mom-and-pop motel in the center of Westcliffe, and this motel was typical of those I frequented throughout my exploration career. I usually wasn't near any large towns that had chain motels, except for an occasional Travelodge. Many of the small towns in out-of-the-way places had only one, or possibly two, nondescript motels, and there would sometimes be other exploration geologists staying there too. Since they were the competition, I tried not to let them know where I was prospecting. We were always cordial toward one another but pretty tight-lipped about what we were working on.

One might think that being alone in so many small towns might lead to spending a lot of time at the local bars. That was not the case, however, and I never knew of any geologists who had this problem, although there must have been some. I suspect that by the end of a long day of hiking through the mountains and hauling out rock samples, we geologists were just too tired to spend much time in a bar.

About this same time, I started a gold-reconnaissance program for Noranda in Colorado. In 1858, placer gold had been discovered in the South Platte River near Denver by prospectors headed to the California gold rush. Our premise was that surely the old-timers didn't find all the gold in the state, so I began an evaluation of the precious metals potential of the San Juan Mountains of southwest Colorado.

The San Juans, as they are informally called, are a huge pile of volcanic rocks about 10–40 million years old with numerous calderas having veins of precious metals (gold and silver). A caldera is a steep-walled, circular structure that forms when a volcanic peak or vent collapses into its emptied magma chamber below. You could think of it as decapitation by collapse. During a thirty-million-year period in the San Juans, explosive volcanism burst forth from numerous volcanic vents with almost unheard of outpourings of ash and lava that coalesced into a shield-like volcanic field a hundred miles across. As the volcanic vents collapsed in on themselves, the calderas were created.

With dimensions of thirty by fifty miles across, the La Garita Caldera is huge, having been produced by the largest volcanic eruption in Earth's history. It released almost 1,200 cubic miles of volcanic ash in one eruption. To put this in perspective, the 1980 eruption of Mount St. Helens yielded only 0.3 miles of ash. Over the next couple of years, I spent a lot of time in this caldera and others; this was fine with me since I discovered they contain a rich tapestry of history and scenery as well as geology.

One of the first exploration targets we identified in the Colorado reconnaissance program was the Mammoth-Revenue gold-silver vein in the Platoro Caldera of the southeastern part of the San Juans. Steve Zahony ended up drilling there, and I gave him help with data evaluation and map compilation. Another target was the old Vulcan copper-zinc-gold-silver mine in the Gunnison area. Sulfur had been mined there in the early 1900s, and copper was produced during World War I. I was not interested in the sulfur or copper potential, but I was intrigued by the possibility of discovering mineable amounts of gold and silver. Bob Gilmore and I ended up staking a few mining claims there in the beautiful but crisp fall weather.

I had a fitting end of the year when I returned to New Mexico to join Bob Hodder on part of the University of Western

Ontario's annual geology field trip to the southwest. I met Bob and his students in Silver City where we toured the Chino and Tyrone open-pit copper mines. From there we drove to Truth or Consequences, where I gave the students a tour of my thesis area in the Salado Mountains. I enjoyed returning to my thesis area and giving back to this new crop of students.

I had been working as a full-time, professional exploration geologist for only three months and already had participated in four different projects exploring for gold, silver, copper, and niobium. My learning curve was steep, I was being intellectually challenged, and every day seemed to bring a new adventure. My confidence was high because I knew I was making a contribution to Noranda.

6. MENTORS BROKEN

My bosses must have quickly developed trust and faith in my geologic abilities as well as in my project management skills because early in 1975 they gave me the added responsibility of conducting a New Mexico molybdenum reconnaissance program. That meant I would now be working in the Rocky Mountains as well as the Colorado Plateau of Utah, Colorado, and New Mexico and exploring for an additional ore mineral. I was being transformed from a geologist who mostly did what he was told into a geologist who led projects on his own. This added responsibility was exciting, and I liked it. And I got a raise!

The New Mexico molybdenum project took me to and from the Ruidoso area in the central part of the state off and on for much of the year. There, while climbing through the Sierra Blanca Mountains, I discovered the Cone Peak and Nogal Peak molybdenum prospects. Steve Zahony and I had to slog through a foot of snow to get to the Cone Peak rhyolite (a fine-grained, light-colored igneous rock) that contained the molybdenum. After further investigation, we were sufficiently encouraged that we staked several mining claims covering an attractive-looking breccia. A breccia consists of angular rock fragments locked in a finer-grained rock matrix; this one contained the ore mineral molybdenite (MoS_2). You could consider a breccia to be the rock equivalent of a gelatin fruit salad.

That snow at Cone Peak did, however, provide me with a good day of fine powder-skiing at Ski Apache on Sierra Blanca Peak. At an elevation slightly less than twelve thousand feet—the highest point in southern New Mexico—Sierra Blanca Peak offered 360 degree views from the top. The ski area is developed on the slopes of a thirty-five-million-year-old eroded volcano that at one time produced violent eruptions. Geology was responsible for this excellent skiing venue.

In early March, all of Noranda was shocked to learn of a helicopter crash in the Panamint Mountains near Death Valley, California. On board were my two mentors, Geoff Snow and Bob Hodder, as well as Tom Evans, the district manager of Noranda's Reno exploration office. A mechanical malfunction of the helicopter had caused the crash, and Geoff, Bob, and Tom all had suffered broken backs. The pilot sustained a serious head injury from which he later died.

A few days after the crash, Steve Zahony and I drove hurriedly through Utah and Nevada to Ridgecrest, California, where the crash victims were hospitalized. I didn't know Tom Evans, but Geoff and Bob were especially important to me because of their solid support and encouragement of my geologic career. They were also personal friends, and I was very concerned about their well-being. Upon walking into their hospital rooms with worry on our faces, all three patients smiled widely and were ecstatic to see Steve and me. We high-fived each other, and Steve and I soon felt much better when we learned all were doing well under the circumstances.

This frightening crash event certainly brought home one of the inherent dangers in this industry—helicopter flying. It didn't prevent me from flying in them, but it did reinforce my attention to helicopter safety. Whenever I flew in a helicopter after that, I always ensured that we were using a reputable flying service and that the pilot had thoroughly conducted his preflight helicopter inspection before we took off.

Steve and I spent a few days keeping our injured colleagues company, and then I drove Bob Hodder to the Los Angeles airport so he could fly to London, Ontario. His family was eager to have him home. After returning to Ridgecrest and ensuring that the other patients were doing well, I picked up Steve, and we began driving home. On this trip, I learned another important lesson: just because two highways intersect in the middle of Utah, it doesn't mean there will be a gas station there. Steve and I almost ran out of gas.

We did not take the direct route home, however. After leaving Ridgecrest, Steve and I took a geologic sightseeing excursion to examine geologic features of interest in Southern California. This trip, like most others, made me feel like Charles Kuralt with my own *On the Road* episodes focusing on geology. Someone once said that the best geologists have seen the most rocks, and I was determined to see as many as I could.

Our first stop was to see our old friend Allen Sphar who was living not far away in Trona, California, on the western edge of Searles Lake in the Mojave Desert. We had a good time catching up on each other's lives. Since Allen was working at Searles Lake, he told us about this intermittently dry lakebed, called a playa, from which the industrial mineral trona—better known as soda ash— is mined. This sodium carbonate mineral is virtually unknown to most people, but it has important applications in the soap-making, glass-manufacturing, paper, and chemical industries; it is a key ingredient in baking soda.

Steve and I visited Kerr-McGee's trona plant nearby before moving on to the Mountain Pass rare-earth mine north of Baker, California. At that time, Mountain Pass was the world's largest producer of the so-called rare-earth elements (cerium, yttrium, neodymium, and others; there are seventeen in all) that have high-tech applications in solid state electronics, compact discs, and magnetic storage media. Magnets made from rare-earths are more powerful than conventional magnets and weigh less, and this is one reason electronic devices have become so small. Rare-earths are essential to a host of green machines, including hybrid cars and wind turbines.

Bastnaesite, the primary ore mineral being mined at Mountain Pass, exists in high concentrations, up to 10 percent; most rare-earth mines extract ore containing only 1 percent rare-earths. The ore is hosted by carbonatite, the same rock type we were exploring at Gem Park in Colorado, so this geologic stop was especially enlightening. We were fortunate that the mine geologist readily agreed to take time away from his duties to show us the geology at this mining operation. A mine geologist is invariably proud of "his mine" and loves to show it off. As an always eager and attentive mine visitor, I was never bored with mine tours.

We made a short stop at the nearby nondescript town of Baker that provided services and housing for the nearby mine. Far from being nondescript today, however, the town is highly memorable because in 1991 a 134-foot-tall outdoor "thermometer" was constructed in the town center; it is visible for miles while driving on Interstate 15. But when it is 115 degrees or more outside, who wants to be reminded of the temperature? This "World's Tallest Thermometer" was built to this height to commemorate the hottest temperature ever recorded in the *world*—134 degrees on July 10, 1913, at Furnace Creek in nearby Death Valley. Although not really a thermometer, it resembles one, illuminated as it is with thousands of lights. Just before arriving in Baker, I was entertained and puzzled by a highway sign announcing the exit for Zzyzx. Who came up with the name of that town, and how in the world do you pronounce it? It rhymes with Isaac's.

This was my first visit to the Mojave Desert, and as I stood atop an outcrop and looked around in all directions at the treeless terrain, I was fascinated by the vastness and emptiness of the countryside—so stark, so naked, so exposed. There was so much open space one could suffer from a serious case of reverse claustrophobia, especially if you were a guy from tree-covered New Hampshire. Only the Joshua trees broke up the barrenness of the scenery. This is the only place in the world where these trees grow, and they are as iconic to the Mojave Desert as the sequoias are to the High Sierra or the redwoods to California's Coast Range.

Steve and I then took a geologic excursion down into Death Valley so we could see for ourselves this hottest place on earth. At 282 feet below sea level, the few buildings identified on the map as Badwater lie at the lowest point in the western hemisphere. My twin brother, Bob, told me that he had once flown as a passenger in a private plane below sea level in Death Valley. I suspect that for a pilot this is a daring must-do sort of thing.

The geology exposed in Death Valley is a collection of colors and various rock types. We saw sand dunes, lots of multicolored volcanic rocks, and mostly barren countryside, all somewhat distorted by the ripples of the emanating heat waves. With its high temperatures and an average of only two inches of rain a year, how did the twenty-mule-team wagon trains and the miners ever survive the summers in Death Valley? We were happy not to be here—in the driest, hottest, and lowest spot in North America— in the summer. Others were not so fortunate; Death Valley was named in 1850 by near-starving survivors of the Bennett-Arcane party of emigrants who lost family members there on their way to California's gold.

Death Valley is framed by the Panamint Range on the west and the Funeral Mountains on the east. These mountains jump right up out of the basin, steep and bare, with no hesitation about it; they don't bother with foothills. From the valley floor, adjacent Telescope Peak in the Panamint Range rises 11,300 feet above the salt flats. Steve and I could see the salt below and the snow above on the peaks, two strips of white with an eleven-thousand-foot wall of barren, brown desert in between.

From the Furnace Mountains on Death Valley's eastern border, I could see snowy Mount Whitney off in the distance to the west in California's Sierra Nevada. At 14,505 feet, Mount Whitney is the highest point in the Lower 48 states. Isn't it fascinating that the highest and lowest points in the continental United States are only eighty-five miles apart? Once again, this phenomenon is a result of geology in action. Starting about five million years ago, nearly continuous movement (in the framework of geologic time) along a series of north-south-trending faults uplifted the Sierra and down-dropped Death Valley by almost three miles. Over millions of years, movement along this army of faults jolted the high Sierra up, earthquake by earthquake, and the range is still growing, as evidenced by earthquakes still occurring there.

The largest earthquake to strike this area in recorded history was the Owens Valley earthquake of March 26, 1872. With its epicenter near the town of Lone Pine at the base of Mt. Whitney, the mountains rose six feet and shifted north fifteen feet. The twelve-foot scarp (a cliff formed by fault movement) is still visible west of town, and it is reported that John Muir felt the jarring of the earthquake 110 miles away in his cabin in Yosemite National Park.

From Death Valley, Steve and I took turns driving as we sped through Barstow and Needles, California, on the way to the Grand Canyon, that 227-mile-long titanic gash in the face of the Earth. I had never seen it before from the ground; what a tremendous exposure of geologic time presents itself there. Looking down at the canyon walls stretching nearly a mile to the bottom, I imagined in the layered rocks periods of time when sandy beaches thrived, sea shells fell to the ocean bottom to become limestone, fine muds settled out of the water to create shale beds, and desert sands accumulated as sand dunes. All these geologic features can be seen in the rocks if one only takes the time to look. Even though I did not hike down into the canyon, I could see that it is one of nature's masterpieces—a textbook of geology waiting to instruct us.

The nearly 2 billion years of Earth's history exposed in the Grand Canyon are represented by rocks ranging from the oldest Precambrian eon at the very bottom up through the younger Paleozoic era at the canyon's rim. The geologic history of the Earth is divided into four distinct lengths of time. The oldest, the Precambrian eon (before life, 542–4,600 million years ago) is further divided into three eras, which I will not burden you with. Younger than the Precambrian eon are three eras: the Paleozoic era (early life, 251–542 million years ago), the Mesozoic era (middle life, 65–251 million years ago) and the Cenozoic era (late life, 0–65 million years ago). Note that the Precambrian constitutes nearly 80 percent of all geologic time. These divisions of time were established in the nineteenth century and were designated according to the fossils found within the rocks. Life evolved from simple to complex organisms; for example, the Precambrian saw the rise of bacteria, the Paleozoic was the age of fish, the Mesozoic hosted dinosaurs, and the Cenozoic saw mammals evolve.

From the Grand Canyon, we passed through the Painted Desert and Monument Valley of northeastern Arizona, two enchantingly beautiful areas where erosion has exposed a kaleidoscope of multicolored sedimentary rocks to everyone's view. Along the way, I drove right into Peabody Coal Company's Black Mesa mine near Kayenta, on the Navajo Indian Reservation, and collected a sample of coal for my personal collection. No one seemed to care since no one intercepted me. This strip mine was the largest mine in the country at the time, extracting coal from the Cretaceous Dakota Sandstone. Standing in the mine, I could imagine the vast expanse of peat bogs, the precursor to coal, which must have thrived here 65–145 million years ago in a hot and steamy subtropical climate adjacent to a beach. Although the multicolored sedimentary rocks of this part of the Colorado Plateau are stunning, the countryside today certainly does not look like a beach.

This had been a worthwhile trip, both in providing emotional support to our helicopter-crash friends as well as in providing comfort to ourselves. And we had participated in another episode of lifelong learning about geologic environments. However, when back in Colorado, the time came to focus on a single geologic environment, one that represented a lone event in the vast history of geologic time—emplacement of niobium at Gem Park. I initiated a program of core drilling and sampling in search of the mineral pyrochlore ($NaCaNb_2O_6F$), which contains the important alloy metal niobium (chemical symbol Nb). Just to confuse things, this element is sometimes called columbium.

An advantage to working out of Noranda's Denver office was its close proximity to the US Geological Survey (USGS) office in Lakewood, Colorado, a suburb of Denver. I commonly went there to talk with USGS geologists who were doing geologic mapping and conducting research in areas that interested us. I had many fascinating and educational conversations with Survey geologists, particularly Tom Steven and Peter Lipman—they helped me tremendously to understand the geology of carbonatites, among other things. Typically, I spent the winter months in the office doing literature research and data compilation, and the USGS geologists were always willing to share their knowledge and expertise.

At Gem Park that spring, we drilled in the snow. Life is not always a geologic excursion to sunny southern California for exploration geologists. Thank heavens for outdoor space heaters called salamanders. Every drill rig had at least one of these diesel-burning heaters that seemed to yield more smoke than heat. But at least the drillers and I could attempt to keep warm. Drill rigs are noisy, dusty, smoky, and smelly affairs—drillers and drillers' helpers have a tough job. They and anyone else who gets near the rigs are always dirty, primarily from the dust and ever-present hydraulic fluid that always leaks from the rigs.

I was assisted at Gem Park by my newly arrived geologic field assistant, Dave Beaty. Dave was a recent Dartmouth graduate from Boulder, Colorado, and he was the first geologist for whom I had supervisory responsibilities. Fortunately, Dave was a pleasant, eager, and smart new hire, so my first supervisory job was not difficult.

Between March and early July, Dave and I supervised the drilling of five exploratory holes at Gem Park. Through geologic logging of drill core, we found that the geology consists of a Cambrian (488–542 million years old), layered, mafic-ultramafic body (dark-colored igneous rock consisting mostly of iron and magnesium minerals) that is intruded by carbonatite dikes (igneous rocks consisting primarily of calcium and magnesium carbonate minerals.) The ore mineral pyrochlore is disseminated within the carbonatite dikes that cut the funnel-shaped mafic-ultramafic body. Carbonatites are rare features in the geologic world, so it was fascinating to study this unusual geology as the core came out of each drill hole.

Layered mafic-ultramafic intrusions are not common either, so I felt *really* fortunate to be working in this geologic environment. This type of geology offers unusual rock and mineral names, such as jacupirangite, okaite, fenite, pyroxenite, lamprophyre, and riebeckite. Many geologists cannot pronounce these names either. I feel blessed to have pyrochlore (the niobium-containing mineral) specimens in my collection. The best known niobium deposit in the world is the Oka carbonatite in Quebec, just west of Montreal.

This was an active mine at the time I was working at Gem Park, but I didn't have an opportunity to visit it, unfortunately.

Dave and I continued geologic mapping and drilling at Gem Park off and on for a few months as we tested our geologic model. We optimistically postulated the existence of a large, buried mass of carbonatite at depth, rich in niobium. Upon the completion of drilling, we concluded that part of our model was indeed accurate; there was more niobium to be found, and we were able to increase the previously identified niobium reserves. That was the good news. The bad news was that the niobium mineralization was erratically distributed, and the host carbonatite dikes were not continuous. There was no large mass of ore-bearing rock, and our geologic model did not hold up in its entirety. No mineable ore was present. Consequently, and sadly, by the end of October, Noranda dropped its interest in the property when our economic evaluation failed to justify further work.

While drilling at Gem Park, Dave and I made a few quick trips to New Mexico as part of my molybdenum reconnaissance program there. We scoured the hills in the central and northern parts of the state but did not find anything worth pursuing. In early April, we drove from Albuquerque, New Mexico, to Moab, Utah, to begin our second drilling program, this one at Lisbon Valley. This trip took us west from Albuquerque to the Colorado Plateau, which for a geologist is a treasure trove of amazing sights. The first geologic point of interest was Grants, New Mexico, where uranium was being mined from flat-lying sandstones by both underground and open-pit methods. This was during the Cold War, so there was strong demand for uranium for use in nuclear weapons. Nuclear power plants were being built also, which furthered the demand for uranium. At the time, this area was one of the country's most important uranium-mining districts.

Beyond Grants, we stopped in Gallup, New Mexico, a classic western tourist town on legendary Route 66. During the 1930s–40s the town was a gathering place for movie stars like Ronald Reagan and Kirk Douglass who were taking advantage of the ready access to western (geologic) landscapes to film movies.

From Gallup, we turned north to Shiprock, a famous desert landmark that is a textbook example of a type of volcanic feature called a diatreme. It is the solidified remnant of the magmatic plumbing system that produced a volcanic neck with a radiating system of magma-related conduits called dikes. The vertical dikes (tabular igneous intrusions that cut through preexisting rocks) stand up in bold relief like walls crossing the flat desert landscape. At its highest point, Shiprock stands seventeen hundred feet above Navajo land and is visible from many miles away.

Hovenweep National Monument and the Aneth oil field came next along this route. Hovenweep preserves six prehistoric Indian villages on mesas and within canyons in southeastern Utah near the Hatch Trading Post. The Aneth oil field is in the Utah portion of the Paradox Basin, not far from the Four Corners where the states of Colorado, Utah, Arizona, and New Mexico come together—of course I had to have my picture taken while straddling those four states. The Paradox Basin was fascinating to me because of its immense thickness (up to ten thousand feet) of salt that had been deposited from an evaporating sea 320 million years ago. I never did see the salt beds, but I read about them in US Geological Survey reports of the area.

We saw lots of American Indians in the Four Corners area, many of them selling their silver and turquoise jewelry beside the highways. I was intrigued by the fact that no silver mines are located on Indian lands; instead, most of their silver for jewelry making was obtained by melting down silver coins and silverware.

After driving farther north, we finally arrived at Lisbon Valley, the site of my next drilling program. This project was a new and exciting challenge because I did not have much experience working with sedimentary rocks. Here was an opportunity to learn, and learn I did. Lisbon Valley sits on the Colorado Plateau, the vast physiographic province that encompasses 130,000 square miles of parts of the Four Corners states: Utah, Colorado, New Mexico, and Arizona. This high, semiarid region is actually a gigantic basin that was filled with sedimentary rocks and then uplifted to elevations ranging from four thousand to eleven thousand feet.

The Colorado Plateau is a wonderful showcase of sedimentary rocks, both for the geologist and the layperson. Looking at these rocks to reveal their secrets is like leafing through the pages of a book. Rocks, like books, shed important light on Earth's history. Both must be opened before their cache of information can be revealed. One requires only the turning of a page; the other often depends on physical means, such as chipping a small rock fragment with a geologist's hammer.

As I examined the oldest layer of rocks at the base of a cliff face one day—the first pages of the book—I saw a thick sequence of marine (ocean) deposits. This indicated that a vast, shallow sea once covered the area. While doing a literature search later, I learned that this thick sedimentary sequence was deposited during the long Paleozoic era of geologic time (251–542 million years ago). The next pages of the book, the younger, overlying sandstones and siltstones, told me that during that period of time (the Mesozoic era, 65–251 million years ago) a floodplain was being created as fine- and coarse grained sediments were deposited as a river overflowed its banks. I saw another layer of rocks that are probably the best known to the layperson—the silty products of swamps and inland seas that contain dinosaur fossils. Isn't every child fascinated by dinosaurs? And I still was.

However, being from the East Coast, I was especially intrigued when I picked up a rock that looked like part of a sand dune. The individual sand grains had a story to tell. I was aware that by examining these sand grains under a microscope, geologists from the US Geological Survey had been able to determine that they were derived from far to the east. This rock sample was part of the 145–200-million-year–old (Jurassic) Navajo Sandstone, a thick stack of desert sand dunes derived from erosion of the Appalachian Mountains, hundreds of miles away. When I picked up that dune-like sample of rock, what an exciting moment it was for me; the sand grains originated from back in my part of the world.

Because of the variety of colors presented by these sedimentary rocks and the effectiveness of erosion, the Colorado Plateau is studded with scenic and unusual landforms that are preserved in well-known national parks—Canyonlands, Arches, Capitol Reef, Bryce Canyon, Zion, Grand Canyon, and Mesa Verde—as well as national monuments—Rainbow Bridge, Canyon de Chelly, and Natural Bridges. We can thank geology again for presenting these stunning pages from the geologic textbook.

At Lisbon Valley, Dave and I explored areas along the Lisbon Valley fault where disseminated and fracture-controlled malachite, azurite (both copper carbonate minerals), and chalcocite (a copper sulfide mineral) were found in sandstones of the Cretaceous Dakota and Burro Canyon formations. The Lisbon Valley fault served as the conduit for ascending copper-rich fluids from which the copper minerals precipitated, and this was our exploration target. In the past, a number of small pits had been excavated to mine the oxide copper to depths of one hundred to one hundred fifty feet. From these pits, hundreds of thousands of tons of ore had been mined. Sulfide copper was present at greater depths below the water table but had not been mined; this was our exploration target.

In addition to our duties supervising the drilling, Dave and I geologically mapped the property and collected rock samples for analysis. Just for fun, we also collected brightly colored blue and green concretions of the copper minerals azurite and malachite. Up to an inch in diameter, these balls weathered out of the rocks and were strewn haphazardly on the ground like a collection of marbles. We also visited nearby uranium properties. Dave's father was a professor at the University of Colorado and apparently was appalled when Dave told him I was carrying a sample of pitchblende (uranium ore) around in my shirt pocket. He was afraid the uranium's radiation would harm me. I didn't have the pitchblende in my pocket for long, so there was no danger.

By the end of this episode of drilling, we had completed seventy-five rotary holes (a rotary drill grinds up the rock and brings it to the surface as small rock chips) around the previously excavated Centennial Pit and proved up (identified) 3,505,000 tons of ore. Combined with the reserves previously identified by

Centennial Development Company, the owner of the property, we now had a total of 9.8 million tons of ore averaging 0.7 percent copper. Not bad work, but this was far short of our objective of twenty-five million tons of 0.6 percent copper. We would have to decide what to do next.

For the Lisbon Valley project, our home base was Moab, a friendly town lying adjacent to the Colorado River. Since Noranda had a relatively long-term presence there, we wanted to be good corporate citizens, so we sponsored a youth softball team, complete with Noranda T-shirts. By getting involved in activities like this, many towns such as this across the west came to feel almost like home after I had stayed in them for a while.

During one interlude away from Lisbon Valley, Dave and I went to Questa, New Mexico, to tour Molycorp's open-pit molybdenum mine with Steve Zahony. We wanted to learn about the geology at the mine because Noranda was exploring the nearby Goat Hill molybdenum prospect farther down the Red River Canyon. Molybdenum occurs in this area in a ring of intrusions poking through the southern rim of the Questa caldera, and Goat Hill was one such intrusion. Back at our motel that night, Steve and I verbally compared mental notes on what we had learned by surreptitiously looking at Molycorp's geologic maps that we had seen posted on the walls of the mine geologist's office. Dave was aghast at the industrial espionage we had conducted. I'm sorry to report that we didn't learn much from this stealth effort, however, and the maps were on public display, after all.

That summer, to cover all my Colorado, New Mexico, and Utah projects, I spent a lot of time on airplanes flying to Moab, Grand Junction, Albuquerque, and Alamagordo, New Mexico. One day in July, I flew from Moab to Denver, spent a few hours with my roommates at the house we rented, and then flew that evening to Alamagordo and drove to Ruidoso, New Mexico. I was young and single, and this was fun and exciting. More importantly, my learning curve at the time was as steep as it possibly could be.

In August, I returned to the San Juan Mountains of southwest Colorado to conduct more gold exploration. One day, I flew from

Denver to Durango and then took a helicopter flight with fellow Noranda geologist John Cleary to examine gold-silver veins high up in the mountains at Ice Lake Basin. Because of some of the rather crude and outrageous things he did or said while in college, John was affectionately known to his closest Dartmouth friends as "Dr. Disgusto." The "disgusting" part was all for show, however, because he is a decent guy and a good geologist.

Blasting molybdenum ore at the Questa
mine in northern New Mexico, 1975.

John and I were primarily interested in the Lucy-Last Hope vein, which had been a source of gold and silver from 1896 to 1917. Even by mid-August—at an elevation of around 12,800 feet—many of the old mine workings (a general term that refers to any underground excavation, such as a tunnel, shaft, or adit) in Ice Lake Basin were inaccessible because of snow cover remaining from the previous winter. This was amazingly beautiful country set among tarns (glacial lakes) and steep-walled cirques, mostly above tree line.

After collecting samples from the Lucy-Last Hope vein for analysis, we let the helicopter go, and the next day we took John's vehicle and drove to the old, abandoned Bandora silver mine located over the next ridge. We bounced around on tortuous and narrow, old mining roads cut into the steep slopes, but the ride was worth the bumps on my head because I collected some wonderful silver-mineral specimens. I could hardly believe I was being paid to go to such beautiful places, and I was being paid to learn so much about geology, too.

This trip was especially unusual because I did all the fieldwork in my street shoes. It turns out I hadn't gone to the Denver airport in a timely fashion, so my baggage, including field boots, did not arrive in Durango with me. Another lesson for a geologist: get to the airport on time.

These two properties we looked at were close to the town of Silverton, so John and I made a quick stop at a rock shop where I bought some beautiful gold specimens. What is a geologist doing *buying* gold specimens? I didn't commonly do that, but these specimens were too good to pass up. They came from one of the first gold mines to see production (1874) in the San Juans, where gold-quartz veins filled fractures in volcanic rocks. Silverton is most famous, however, as an old silver-mining town lying in a beautiful and remote location at an elevation of 9,308 feet with thirteen-thousand-foot peaks surrounding it. Fewer than five hundred hardy folks can stand its harsh winters.

From Silverton I drove by myself to northeast of Gunnison where I hiked into Lamphier Lake in the Sawatch Range with the intention of camping out. This was really high country, containing three of the five tallest mountains in the Lower 48 states. The range is a large, faulted anticline (an upward fold in sedimentary rocks) intruded by Tertiary igneous rocks, and those igneous rocks were what I was seeking. The exploration target was the Broncho Mountain molybdenum prospect that had been brought to Noranda's attention by a prospector.

As was always the case, my constant companion out there in the wilds that day was wonder—wonder at what I would find and experience around the next ridge. What mystery would I find? And what mystery will I unravel? My geologist's insatiable curiosity is what kept me going. I camped for the night in a tent next to a tarn just below tree line, and the next day I examined the prospect. The hike was great but not so the prospect. I was impressed, however, with the abundance of those blue-green tarns nestled in the cirques.

I also looked at the Bessie G gold-silver vein in the La Plata Mountains. Surprisingly, I discovered that even after many years of mining, abundant free gold could still be found there. (Free gold is native gold or pure gold, rather than a chemical combination of gold and another element.) What the old-timers said was true, "There's gold in them thar hills." The La Plata Mountains form the rugged southwestern margin of the San Juan massif (a large mountain mass) and are clearly visible from miles away. They are the last mountains of the Colorado Rockies before heading off to the flat Colorado Plateau to the west. After discovery of gold there in 1873, these mountains eventually produced a quarter of a million ounces of this precious metal. I, on the other hand, did not discover any worth mining.

About this time, I returned to the Vulcan property south of Gunnison where Dave and I conducted a surface geophysical survey of the property. I don't recall what kind of geophysical survey it was, but it must not have been too difficult because neither Dave nor I knew much about the arcane field of geophysics. While Dave was away one day, I spray-painted "Dave's Place" on the side of an old miner's cabin at the abandoned town site of Iris. I thought it was funny at the time, but I now regret defiling that old cabin with graffiti. When I see graffiti today, I am reminded of this regrettable act.

While in southwest Colorado, Dave and I made a visit to yet another national park, Mesa Verde. I was fascinated by the cave-sheltered dwellings of the American Indians who inhabited this area from AD 900 to AD 1250. Did you know that geology is responsible for this superb village setting? Impermeable beds of shale lying beneath a porous sandstone unit cause ground water from the sandstone to flow along the top of the shale. Weakened by seepage from small springs where the ground water emerges at a cliff face, the overlying sandstone falls away in large arcs to create the arched caves that provide shelter. In this cliff-dwelling setting, the American Indians were thus provided with protection from the elements.

Earlier in the summer, I had taken a few days to attend a meeting in Silver City, New Mexico, titled "Conference on Base and Precious Metals Districts of New Mexico and Arizona." This meeting provided an opportunity for exploration geologists and academics to meet and share their knowledge about precious metals. Events such as this helped me keep up with exploration that was taking place in the two states as well as with the latest geologic research regarding ore deposits. The meeting was sponsored by the New Mexico Geological Society who later that year published in their fall field conference guidebook a technical article I had written, "A Model for Subduction Origin and Distribution of Fluorite Deposits in the Western United States." What a thrill it was to see my name in print in a professional publication. I could see that my stature in the industry was indeed increasing.

In mid-December, I flew to London, Ontario, where in collaboration with Bob Hodder, I wrote another technical paper on fluorite metallogeny (how and where fluorite occurs in the Earth's crust), "Distribution and Genesis of Fluorite Deposits in the Western United States and Their Significance to Metallogeny." This was an elaboration of the main outcome of my master's thesis research. I felt thrilled to be writing a professional paper with a distinguished university professor. I was especially proud when a few months later the paper was published by the Geological Society of America in their monthly magazine *Geology*. I was amazed at how much progress I had made in my professional development. A few short years earlier, I would not have thought that I would ever be a published author.

7. FIRST LOVE–THOMPSON CREEK

Geoff Snow, the president of Noranda Exploration, believed strongly in ongoing education for the company's geologists. Because I, too, valued continuous learning, you can imagine my excitement when I learned that Geoff had planned a company field trip to Central America. He was taking us to the country of Guatemala to examine young volcanic rocks and volcanic processes that could lead to ore deposits. I had been to Guatemala twice as an undergraduate geology student at Dartmouth, but I was only a fledgling geologist then; I now knew I had missed a lot on those trips, and I was determined to make up for it.

Guatemala is a highly instructive place for geologists because of the easy access to its chain of active volcanoes paralleling the Pacific Ocean coastline. Older volcanic rocks in many parts of the world commonly contain ore deposits, and these modern-day Guatemalan volcanoes provided an opportunity for me to improve my ability to understand the ore forming processes. Volcanologically, Guatemala is an active place, and I saw firsthand superb examples of stratovolcanoes, ignimbrites (ash-flow deposits), pumice deposits, lava flows, fumaroles, volcanic domes, lahars (volcanic mud flows), and, of course, eruptions of ash and lava. These geologic phenomena were just like the textbooks described them, only more dramatic and memorable. With the guidance of the field trip leaders, I learned to appreciate how these features could lead us to ore deposits.

Twenty-five Noranda geologists flew to Guatemala City on January 15, 1976, and spent the first night at the Pension Asturias, a small hotel where Dartmouth's third-year geology majors stayed for two weeks each fall while learning about volcanology. I remember my time spent with twenty other young geology students in the fall of 1970 as being a wonderful hands-on field experience, but we did not learn much about ore deposits. I also remember being slightly overwhelmed then by the strange culture surrounding me. This trip with Noranda was a return to familiar country but with a different geologic focus.

That night at the Pension Asturias, we were honored to meet our distinguished cast of field trip leaders: Howel Williams, America's informal dean of volcanology and emeritus geology professor from the University of California at Berkeley; Dick Stoiber, the preeminent expert on Central American volcanoes and geology professor at Dartmouth College; my mentor Bob Hodder, an expert on the relationships between volcanic processes and ore deposits, and geology professor at the University of Western Ontario; and Sam Bonis, a geologist with the Instituto Geografico Nacional de Guatemala. Alfredo MacKenney, a medical doctor in Guatemala City, joined us for one day on the hike to Pacaya volcano; he is a volcanophile who climbs erupting volcanoes just for fun. The active vent on Pacaya volcano, called the MacKenney cone, is named for him.

On our first full day in Guatemala, we Noranda geologists climbed Pacaya volcano, only twenty-five miles south of Guatemala City; we could see it erupt from our pension and from the airplane as we landed the day before. While climbing up the steep slope of the volcano by slowly putting one foot in front of the other on loose, gray cinders deposited from a recent eruption, I could see above me volcanic ash and bombs (masses of semisolid magma) being ejected high into the sky from Pacaya's crater. Being daredevils, once at the crater rim, some of us tried to capture good photos of each other standing there bravely (foolishly?) with the eruption billowing over our heads. With adrenaline coursing through my veins in the heat, noise, smell, and dust of the eruption, I realized this volcanic event occurring before my eyes must be similar to those that had produced Colorado's San Juan Volcanic Field thirty million years earlier.

The summit of Pacaya provides an excellent spot for viewing three other close-by volcanoes of Guatemala's Tertiary volcanic province: Fuego, Acatenango, and Agua. And off to the west were beautiful Pacific Ocean beaches flanked by coffee plantations. The return from Pacaya's summit was fun also, as we glissaded (without skis) down the slopes through clouds of fresh ash.

With the exhilaration of the Pacaya eruption still in our minds, we went to Fuego volcano where we witnessed a small ash eruption from a distance. We did not climb that volcano; we had had enough excitement for one day. We stopped for the night in the old Mayan Indian town of Quetzaltenango, hidden away at 7,655 feet in Guatemala's Central Highlands. This picturesque city preserves the old traditions and architecture of its Spanish colonial past. The city was already some three hundred years old when Spanish conquistadors conquered Guatemala in the early 1500s. Quetzaltenango means "the place of the Quetzal bird," and, true to its name, we enjoyed seeing these brightly colored and noisy birds along the way. The Quetzal is Guatemala's national bird as well as the name of its currency.

Clearly visible only seven miles south of Quetzaltenango was the silhouette of Santiaguito volcano. The next day, we hiked up that volcano where we were greeted firsthand by two small ash eruptions from its Caliente Crater. A group of us was standing only about three hundred yards away from the crater when suddenly sulfurous steam started spewing from cracks in the rocks beneath our feet. This was followed by grayish ash coming from the volcanic crater and falling everywhere, including on me. My head and shoulders were covered with the gray, powdery fluff. Thank heavens it wasn't hot.

Even though these were small eruptions and not particularly dangerous, they certainly were exciting to witness so close at hand. Peering into the many fumaroles, which are vents from which volcanic gases spew forth, was thrilling as well as choking. In addition to sulfur and other noxious gases, metals sometimes accompany these gases, and at depth they may precipitate in sufficient concentrations to constitute ore. That night I watched Santiaguito erupt again, this time from two miles away where we camped at "the casita", basically a lean-to on the lower slopes of the volcano. Sleeping bags with tired bodies were scattered everywhere. As I dozed off, I listened to rocks tumble down the slopes of that fascinating mountain.

One day as we headed west on the Pan-American Highway out of Guatemala City to the country's Central Highlands, we were treated by seeing many cultural sites along the way. Bright colors were everywhere—in the goods being sold in the open-air markets as well as on the Mayan merchants who wear traditional hand-woven, brightly colored clothes. I took lots of pictures of Mayan women carrying all sorts of things on their heads, from vegetables and tortillas to pots of water or bolts of woven cloth.

Especially intriguing were the "chicken buses" on the Pan-American Highway. These old, yellow US school buses are bought in Texas and driven to Guatemala where they are then repainted in bright colors and their gasoline engines replaced with large, powerful Caterpillar diesel engines that belch black smoke. Speed limits are not enforced, so the drivers travel at maximum speeds possible while passing on blind curves and performing other terrifying maneuvers. The buses carry anything and everything on their roofs, including cages of live chickens, hence the name "chicken buses." How colorful!

For relaxation one day, we enjoyed a sun-soaked afternoon at Lake Atitlan, about which German explorer Alexander von Humboldt, the earliest prominent foreigner to see the lake, said, "It's the most beautiful lake in the world." From its 5,125-foot elevation, I enjoyed stunning views of indigenous villages nestled among ten-thousand-feet-high Toliman, Atitlan, and San Pedro volcanoes rising at the water's edge on the south side of the lake. In the far distance to the north, I could see Santa Maria and Santiaguito volcanoes.

The 1,115-foot-deep Lake Atitlan fills a miles-wide caldera (remember, a caldera is a large, sunken volcanic crater that forms when the ground surface collapses after magma has erupted). The nearly vertical north wall of the caldera provides a majestic setting adjacent to the lake, and I have a picture of some of us sitting on the beach talking shop and enjoying the scenery of blue sky with gray volcanoes jutting majestically upward. However, I do not have a picture of fair-skinned Bill Block looking as red as a boiled lobster after sitting in the hot, tropical sun much of the day.

For a cultural enrichment day, all of us geologists flew on a DC-3 airplane to the Tikal Mayan ruins in the rain forest of northern Guatemala. Flying low over the thick jungle was a little scary, especially when suddenly the tops of giant pyramids began poking above the greenery. The Mayan culture thrived here for more than fifteen hundred years, and the Mayans once inhabited much of Central America. With more than four thousand stone structures, Tikal is the most famous and most impressive Mayan archaeological site in Guatemala. The Mayans built pyramids for sacrificial rituals and as places of entombment, and they were built taller than the jungle so they could be used by the citizens as landmarks. Having never seen anything like this before, I marveled that the Mayans were able to build a highly functioning city out of the thick jungle and that they occupied it continuously for fifteen hundred years.

This visit to Guatemala was a wonderful trip in all respects—it provided geologic insights into ore-forming environments and heightened my awareness of cultural differences.

Sadly, one week after returning home, a magnitude 7.5 earthquake struck Guatemala; 2,300 people were killed, and significant structural damage occurred. We were fortunate not to have been there, but had we still been in the country, we possibly could have provided help to the natives. Most adobe houses in the outlying areas were destroyed, and transportation was disrupted for days by landslides.

The earthquake originated about one hundred miles northeast of Guatemala City along the Motagua fault, a fault that we had examined during our field trip. The Motagua fault separates the North American tectonic plate from the Caribbean plate, and the fault movement was horizontal instead of vertical—geologists call it strike-slip motion. This earthquake was caused by the same kind of tectonic movement—that is, horizontal—that produced the 1906 San Francisco earthquake. Guatemalans are a strong people as evidenced by their resilience to frequent volcanic eruptions as well as earthquakes. I learned another sad thing when I returned to Guatemala City in 2009 on a family trip—the part of town around the Pension Asturias had been taken over by gangs and is no longer a safe place.

Once back at work, Noranda decided to drop the Lisbon Valley property, so I spent much of my time writing the final project report. The decision to drop the property was based on economic factors. Under the terms of agreement with Centennial Development Company, the financial rate of return on Noranda's investment in the property would not have been sufficient to justify a mining operation. Copper prices were just too low. However, with a rebound in the price of copper in the late 1990s, another company did put the property into production. I was gratified to hear that.

Dropping the Lisbon Valley property came as a blow to me since I had invested so much time and effort in exploring it, but I understood the financial realities and didn't become discouraged. The Lisbon Valley project had value for me because it had been educational as well as fun—I had gained some expertise in sedimentary rocks and the copper mineralization they sometimes contain. This had been an opportunity that helped me become a more well-rounded geologist.

In April, one of my roommates and I started driving from Denver to California for my twin brother, Bob's, wedding. In Craig, Colorado, we picked up Jeff Stimson, another Dartmouth geologist with whom I had attended grade school in Bath, New Hampshire. We met Bob and his fiancé in Reno and then drove to Grass Valley, California, the site of the wedding.

After the joyous event, I drove back to Denver with my mother and father; I played geologic tour guide the entire way. My parents stayed with me for a while in Denver, and one weekend we visited Rocky Mountain National Park. Another day we drove to the top of Pikes Peak so they could compare it to Mt. Washington in New Hampshire. At 14,115 feet, Pikes Peak is almost eight thousand feet higher than Mt. Washington, New Hampshire's highest peak. If I remember correctly, this was my parent's first trip west, and it was a real adventure for them. I certainly enjoyed showing them the sights, and they appreciated getting a feel for where I was working.

I enjoyed giving my parents the Rocky Mountain tour, but soon it was time for their return to New Hampshire and my return to work.

Noranda became interested in the geology around the town of Creede, Colorado, so a couple of us geologists made a visit there. We found that Homestake Mining Company had beaten us to the punch by staking the most attractive ground just twelve days earlier. That saying about the early bird and the worm comes to mind. We did, however, discover the old Yukon-Alaska precious metals property on the north side of the Creede Caldera east of Silverton, Colorado. After a preliminary examination, it looked attractive. We observed rocks that had been altered by hot water coming up from far below the ground surface and quartz that had precipitated out of those fluids in veins. Alteration of rocks is one of the more pervasive indicators of hydrothermal activity, and it was always a welcome sight. Maybe there was gold and silver in high enough concentrations that mining could be justified. So Bob Gilmore—from the Lisbon Valley project—and I spent twelve-hour days staking claims there. Although it was hard work, we enjoyed the gorgeous weather and beautiful scenery surrounding us.

Later on, I went back to the Ruidoso, New Mexico, area to collect soil samples from the Cone Peak molybdenum prospect that I had discovered last year. I intended to have the samples geochemically analyzed to determine if there were any indications that molybdenum might lie below. I pitched my bright red tent among some tall trees on Big Bear Creek and listened to the rain, hail, sleet, and thunder during the night. Cozy as I was in my down-filled sleeping bag, I awoke the next morning to three inches of fresh, fluffy snow on the ground outside my tent. Although beautiful, it was wet and cold and certainly not conducive to soil sampling. So I packed up my gear, hiked out, and started sampling other properties that were at a lower elevation and not snow covered. After awakening the next morning—in a motel room this time—to find snow on my vehicle, I packed up and flew back to Denver.

I made good use of this unexpected return to home base by taking a week-long course at the Colorado School of Mines titled, "Economic Evaluation and Investment Decision Methods." This was an excellent course that taught me how to evaluate the financial viability of different mining scenarios. I had never been taught anything like this in college, and it served me well during the coming years as I applied what I learned to numerous mineral properties. My repertoire of valuable skills was broadening.

For the last three years, I had spent about 80 percent of my time on the road. Even though that was a long time to be away from home, it was an enjoyable and heady time for me. Having no personal commitments back in Denver to tie me down, I was free to learn about and experience new places and people and investigate a wide variety of geologic settings. Most of the places and geologic environments were new to me, and I soaked up everything I could about them; my learning curve was incredibly steep. Being constantly bombarded by new experiences, I loved every minute of these adventures.

Typical of many campsites, this is the author's tent at the Cone Peak molybdenum prospect in New Mexico's Sierra Blanca Mountains, 1976.

Looking back on it, I realize I had subscribed to a philosophy summarized by Glen Heggstad in his book, *One More Day Everywhere*: "When it comes to adventure travel, you can't take a wrong turn."[1]

I did not know it at the time, of course, but June 1976 was the start of what became the highlight of my exploration career—my first love, if you will. I began a long and wonderful association with the Thompson Creek molybdenum project about thirty miles southwest of Challis, Idaho. Cyprus Mines Corporation of Los Angeles was seeking a joint-venture partner on this molybdenum property where they had extensively explored the surface and underground. Having excavated 8,600 feet of underground mine workings on two levels and drilled numerous diamond drill holes, Cyprus had outlined 110 million tons of ore averaging 0.152 percent molybdenite. This was quite a large tonnage of respectable ore grade, and we all hoped there was more ore to be found. My job was to lead an exploration program to determine if Noranda should form a partnership with Cyprus to mine the deposit. So, on June 1, I flew to Missoula, Montana, and then drove south through Montana's beautiful Bitterroot Valley to Salmon and Challis in the rugged central Idaho mountains.

Glenn Gierczcki—a summer geology student hired by the Denver office—and I rented a cabin at Torrey's Cabins right on the Salmon River near Clayton, Idaho, the closest community to Thompson Creek. Phil and Val Johnson, a nice older couple, were the proprietors, and they enjoyed our company as much as we enjoyed theirs. Glen and I basked in the simple pleasure of barbequing on the outdoor grille for dinner while watching the Salmon River flow by. I wasn't a particularly good chef, but some of the field assistants who joined me later certainly were. Clayton was mostly just a wide spot in the road, but Challis, a few miles downriver with a population of around twelve hundred, was the ranching supply center for the area and the county seat of Custer

---

1.   Glen Heggstad, *One More Day Everywhere, Crossing 50 Borders on the Road to Global Understanding* (Toronto: ECW Press, 2009).

County. Interestingly, a Custer County can also be found in five other western states. That Custer person sure got around.

Sadly, in 1983, Challis took the brunt of a magnitude 7.3 earthquake that killed a first- and second-grader as they walked to school. They were hit by debris falling from a building. The earthquake was named the Borah Peak earthquake because its epicenter was near this 12,662-foot peak in the nearby Lost River Range. After the earthquake, the peak was about eight inches taller than it had been. This earthquake was the strongest ever felt in the state, and sidewalks in faraway Boise could be seen rippling during the event. For those interested in seeing vivid examples of geologic features, the twenty-one-mile-long fault scarp created by the rupture is clearly visible along Highway 93 at the base of the Lost River Range. This impressive fault scarp varies from nearly six to sixteen feet in height.

The strongest earthquake I have experienced is the 1989 Loma Prieta earthquake, which was centered about forty miles from where we lived in Pleasanton, California. Loma Prieta's magnitude was 6.9 on the Richter scale, and because it displaced the Earth's crust by about five feet, its vibrations caused the window shades to rattle soundly at my place of work a few miles east of Pleasanton. My wife, Janet, stood with little Meaghan and a friend of hers beneath the door jam in our house watching the entryway chandelier sway back and forth for twenty seconds. She did not share my enthusiasm for this kind of geology in action! The famous nineteenth-century geologist and charter member of the US Geological Survey, G.K. Gilbert, said, "It is the natural and legitimate ambition of a properly constituted geologist to see a glacier, witness an eruption and feel an earthquake." I have done all three and loved it. Whereas many people would have been frightened, I was enthralled by these examples of Earth's might.

At Thompson Creek, we began a major exploration program with Centennial Development Company as the contractor performing underground mining work for us, and Boyles Brothers Drilling Company providing core drilling services. By the end of the year, we drilled 10,091 feet in five holes cored from the surface and nine holes cored from underground. We kept Centennial Development Company very busy digging underground workings to provide access for the drill rigs. (Using the word "very" reminds me of Geoff Snow's effective use of humor while instructing me in good writing. He once told me that "very" is a useless word; when you're writing, you should cross out "very" and write in "damn" and then cross out "damn.")

The drilling and mining crews were not only extremely competent and professional but fun to be around in non-work hours as well. I spent a lot of off-hours time with Floyd Marshall and his wife; Floyd was the supervisor of the Centennial Development Company crew, and he and his wife were an engaging couple. Being with them and the drilling crew satisfied a need for socialization. One Friday, we all took the afternoon off to attend a high school football game in Challis where Floyd's son was playing. What a fun afternoon after a busy week!

During that year, Glenn and I busily split and geologically logged drill core, supervised the excavation of underground workings, geologically mapped those workings, and generally tried to unravel the complexities of Thompson Creek geology. With a profusion of head scratching, we were giving the rocks an opportunity to speak to us.

I developed work plans and provided daily direction to the work crews to ensure that exploration was carried out effectively and efficiently. Being responsible for the work of four Centennial Development Company miners, four Boyles Brothers Drilling Company drillers, a bulldozer operator, and eventually three geologic field assistants was a challenge. This was the largest number of people I had ever supervised, and through trial and error, I learned what worked and what didn't. I seemed to have an innate sense that by treating people fairly and with respect, I could get the most out of them.

The mine portal (opening to the underground workings) was right beside our small field office, and I always got a thrill when I walked through it into the heart of the orebody. Riding with the miners through the portal in their rubber-tired mucking (mining) machines was also exciting. But working with the miners was *always* fun. One of their favorite tricks to pull on a new, young geologist was to hand him a piece of concrete and ask him what kind of rock it was. The geologist had better get it right or he was in for a lot of ribbing. And the miners could be merciless about their teasing. I successfully passed that test—concrete is a manmade product, not a rock. Early on, however, I did learn a valuable lesson from Geoff Snow about working underground: "When you're talking to someone underground, don't look at them directly in the face, because when you do, the beam from the mine light on your hard hat hits them directly in the eyes." That will brand you as a novice immediately.

Glenn Gierczcki and I conducted surface reconnaissance mapping of the host for the molybdenite, the Cretaceous (eighty-five-million-year-old) Thompson Creek quartz monzonite (a light-colored intrusive rock similar to granite), and found spectacular outcrops of stockwork (a network of crisscrossing) quartz-molybdenite veins. Molybdenite is a beautiful mineral. When fine grained, it is dark gray and resembles graphite; when coarser grained, it is bright, silvery colored. We had both types.

The Thompson Creek stock (a relatively small body of intrusive rock) and its molybdenite probably crystallized from molten magma fifteen miles or so beneath the Earth's surface. You can think of the stock as a fossilized magma chamber. Uplift and erosion over the past eighty-five million years have stripped away much of the overlying, non-ore bearing rocks, thus partially exposing the molybdenite. Thankfully, the ubiquitous forces of erosion—rain, wind, running water, and glaciation—were sufficient to expose the formerly buried orebody to human view. All it took was a geologist wearing hiking boots and a Filson vest to come along and find it.

As Glenn and I were geologically mapping at the far back corner of the property one day, we were surprised to discover an old adit that hadn't seen any attention in years. We knew it was old, because inside were wooden ore-car rails, not metal ones. I recognized these as treasures, so I picked one up as a souvenir.

One Sunday, Glenn and I went geologic touring and sightseeing in the surrounding Challis National Forest and Sawtooth National Recreation Area, the latter of which is high, glaciated country with many alpine lakes. This remote area of ponderosa pines, aspen, and sagebrush is home to elk, deer, cougars, wolves, foxes, bighorn sheep, and eagles. I could see that the Sawtooth Range is aptly named because the serrated ridges at its crest look like the teeth of a saw blade.

This mountain range is only a few miles west of Thompson Creek and presents a dramatic skyline for anyone fortunate enough to see it. The range has been uplifted along a fault and towers over the down-dropped and wide-open valley floor of the adjacent Stanley Basin. Sitting at the base of the Sawtooths are the towns of Stanley (population about forty in winter) and Lower Stanley (population less than forty at any time). These towns hug the beautiful, free-flowing Salmon River, the "River of No Return." The river got its nickname from the fact that farther downstream it flows so swiftly that in the early 1900s it was impossible to bring boats back up the river, so they were dismantled, and the boards were used as lumber. Since these boats never came back to their starting point at Salmon, the "River of No Return" moniker was coined.

Stanley is a hopping place, especially in the summertime when tourists invade the area. Noted for its annual Stanley Stomp, tourists might catch this big party and dance that spills out into the streets. In the winter, however, at an elevation of 6,260 feet and surrounded by high mountains, Stanley is often the officially coldest place in the Lower 48 states.

The stunning Stanley Basin is home to herds of grazing cattle. What a view the cowboys herding those cattle enjoy every day of the glacially carved Sawtooth, White Cloud, and Boulder Mountains surrounding them. The Sawtooths are especially impressive with their eleven-thousand-foot sharp peaks and hundreds of alpine lakes. The water is so crystal clear that I could see ripple marks in the sand at the bottom of the lakes and streams.

A little upstream from Stanley is a fish hatchery on the Salmon River where sockeye and Chinook salmon and steelhead trout swim upstream to spawn. How amazing it was to see these two-foot-long, fifteen- to twenty-pound fish jumping up to five feet in the air as they made their way to their birthplace. Salmon and steelhead trout travel about nine hundred miles from the Pacific Ocean to spawn in the headwaters of the Salmon River, the longest fish migration in the country. The fish swim up the Columbia River past Portland, Oregon, through southern Washington to the Snake River, and then to the Salmon River where they eventually end up near Stanley, where they were born.

This is Idaho Batholith and Sawtooth Batholith country. A batholith is a large body of intrusive rock (larger than a stock) that often is exposed over thousands of square miles. These Cretaceous (65–145 million years old) granitic rocks form the high backbone of the state. The Idaho Batholith stretches two hundred miles northeast from Boise all the way across central Idaho and into Montana, covering 15,400 square miles of mountainous terrain. Most of it is inaccessible because of lack of roads or federal wilderness designation.

Because of the abundance of new-to-us geology in the state, one day Glenn and I embarked on another Charles Kuralt-type of *On the Road* trip. This time we went to Craters of the Moon National Monument just a few miles southwest of the small town of Arco. This vast lunar-like terrain of black basalt lava flows and cinder cones has all the volcanic features imaginable. I thought it looked like the moon, and so did others apparently, because NASA had assigned *Apollo* astronaut Alan Shepard and his crew to train there in 1969. They explored the lava landscape and learned the basics of volcanic geology in preparation for an upcoming moon landing. This short course in geology was necessary because, of all the NASA astronauts, only Harrison Schmitt was a geologist. Specimens of actual moon rocks the astronauts brought back are indistinguishable from the basalts at Craters of the Moon.

The town of Arco itself is famous because in 1951 it was the site of the first peaceful use of nuclear power—four lightbulbs were lit by nuclear-fission-generated electricity. Arco is near the

US Department of Energy's (it was called the Atomic Energy Commission then) Idaho National Laboratory, the site of the electricity-producing nuclear reactor.

The area around Arco, including Craters of the Moon, is part of the vast ocean of lava that created the Snake River Plain of southern Idaho. This lava plain is the product of young eruptions from fissures (cracks in the Earth's crust). Some eruptions were as recent as two thousand years ago, and they produced basalt lavas that flowed freely like honey. This forty- to sixty-mile-wide lava plain forms a wide scar across the face of southern Idaho as it cuts at right angles across the north-south grain of the mountains. You can clearly see this discontinuity on any map.

As Glenn and I drove across the Snake River Plain, I could see no evidence that lying beneath the lava beds is a chain of older volcanic centers that erupted violently millions of years ago—I saw only an abundance of potatoes, evidence that this *is* most certainly the "Famous Potatoes" state. Eruptions from the now-hidden volcanic centers belched ash high into the sky, and when the ash settled on the ground, it cooled and hardened into light-colored rocks called rhyolite tuffs (fine-grained, silica-rich volcanic rocks) that are exposed only in the deepest canyons cut into the plain by the Snake River. Calderas formed by these eruptions were subsequently filled in by the younger basalt flows that we see at the surface. See what Mother Nature hides! Along a five-hundred-mile stretch beneath the Snake River Plain, there are ten calderas, the oldest being 16.5 million years old in southeast Oregon. The youngest are the calderas in Yellowstone National Park (more on that later).

Early in the summer, I flew to Los Angeles to review and pick up geologic data from Cyprus' office headquarters. This was my first trip to Los Angeles, and I was shocked and dismayed by the immensity of this sprawling metropolitan area. People and buildings were everywhere; too many and too confining for me. I had never seen a city so large, and I didn't like it. On the way there, I first flew to Stockton, in central California, and then drove to Sonora in the foothills of the Sierra Nevada. My identical-twin brother, Bob, also a geologist with Noranda, was there mapping a gold prospect, so I spent some time catching up with him and learning about California geology. This part of California I liked better.

Since we are identical twins, it may not be surprising that both Bob and I chose geology as a career. What *is* surprising is that two others from Bath, New Hampshire—Jeff Stimson and Roger Thrall—also became geologists. They were one year behind Bob and me at Bath Elementary School. Equally surprising, all four of us attended Dartmouth College where we majored in geology. To put this unusualness into perspective, my eighth-grade graduating class at Bath Elementary had only eight students. It must be that the rural, small-town environment of Bath encouraged outdoor exploration and sparked our interest in nature.

On July 4, 1976, I celebrated the nation's two hundredth anniversary from the top of Uncompahgre Peak in Colorado's San Juan Mountains. At 14,309 feet, it is the sixth highest peak in the state. Three friends from Denver and I thought it would be fitting to celebrate the occasion by enjoying some of the country's natural splendor, so we backpacked into the mountain and camped overnight. Even on my days off, I was doing something related to geology.

Underground mine geologist Lamarre standing
by a mucking machine at the Thompson Creek
molybdenum deposit, Idaho, 1976.

Back in Idaho, I awoke on August 16 to find fresh snow on the ground at Torrey's Cabins—in August, mind you. The gorgeous early fall weather and scenery in central Idaho reminded me of New Hampshire. The yellow-leafed aspen trees on nearby Railroad Ridge were on full Technicolor display one day as we prospected for more outcrops of molybdenite.

In early September, Asa Clark arrived to replace Glenn Gierczcki, who had returned to school. Asa, a hardworking and good-natured young geologist, grew up on a farm in Pullman, Washington, where his father grew hops in the Palouse Hills for the breweries. Another recent college graduate, Dennis Kenney, arrived shortly thereafter from Florida to assist us. Dennis is a character; when he saw something he liked, he'd often say, "That's better than ten average things." He had opportunities at Thompson Creek to say that frequently. I really appreciated the fact that he so eagerly shared my enthusiasm.

In mid-September, my pickup truck became stuck in the mud (more correctly, I got it stuck in the mud) on one of the drill roads above the mine portal. Because I had to drive to Boise that day to catch a plane, I took another vehicle and left mine stuck there, with instructions for the others to, "Please get it unstuck for me."

I flew to Vancouver and then drove along the Cariboo Highway to the community of 108 Mile House near Kamloops, British Columbia, to attend a meeting of the local section of the Canadian Institute of Mining. After the meeting, Noranda held an in-house symposium on molybdenum deposits; we discussed Noranda's Adanac, Boss Mountain, and Thompson Creek properties, and I was thrilled to lead the discussion of Thompson Creek. How exciting it was to gain insight into these other molybdenum deposits within the Noranda family.

Being a geologist at Thompson Creek was not just about the science of geology. I had technical matters to deal with also. One day the drillers dropped a downhole-survey instrument to the bottom of one of the drill holes. That's an expensive problem because not only is the instrument itself expensive but so is the time required to retrieve it, which the drillers eventually did. My field notes record the observation, "Drillers giving me a headache as usual." That was written in a moment of frustration, because all in all the drillers and mining crew were highly competent and professional.

With a smile, I remember mapping underground with my summer field assistant, Glenn, one day. As three o'clock in the afternoon neared, he finally asked in exasperation, "Aren't we going to have lunch?" I was so enthralled with what I was doing, standing in water up to my knees while examining rocks through my hand lens, that I had lost all track of time. I also often lost track of time back at Torrey's Cabins where we often worked late into the night plotting maps and debating the geology.

I recall a poem Stew Wallace, a well-known Dartmouth geology graduate, introduced me to: "To commune with the rocks is a special privilege; to study beneath the ground is a noble profession." Stew was famous in the exploration community for discovering the Ceresco Ridge orebody at the Climax molybdenum mine as well as the Henderson molybdenum deposit at nearby Red Mountain in Colorado—he's one of the three ore discoverers I know. (The other two are Eric Struhsacker and Dave Jones.) I was honored to have Stew spend a couple of days with me at Thompson Creek, going underground and sharing his invaluable insight about what we had found so far. I felt like I was in the presence of a god. I don't know if the poem he recited is original with Stew or not, but it certainly was appropriate. To thank Stew for an educational and productive session with us, Glen and I took him out on the town in nearby Stanley where we all had a good time.

For me, it was exciting to go into the workings of a newly excavated underground mine after drilling and blasting had been done and know that I was the first human being *ever* to see those rocks. This experience was almost mystical! As I stood amidst a pile of broken rocks, breathing fumes from the recent blasting and examining a rock sample with my hand lens, it was an exciting challenge knowing that in the rock types, mineralogy, hydrothermal alteration, faults, and joints exposed in the newly created rock exposures, all the necessary clues were present to help me understand how and why the ore got there. As the rocks spoke to me, I could unravel their story. What does this rock type mean? What is this mineral? Why is there molybdenite here?

Working in the darkness, stillness, and quiet hundreds of feet underground is not for everyone, but I loved it. Although never frightened or claustrophobic, I was usually dirty and sometimes cold and wet. I went into an abandoned adit in Arizona once and was surprised to find a badger living in its far end. That caused my adrenaline to report for duty. I left the badger alone.

Recently I had started seeing Janet, who was the executive secretary in Noranda's office in Denver. She had been raised on a sheep ranch in Wyoming and had worked for oil and gas and minerals exploration companies, so she was familiar with the natural resource industry. We got along well, and she enjoyed learning more about what geologists do in the field, so in mid-December she flew up to Idaho to see me. I picked her up at the Sun Valley airport. Although she didn't especially like flying in the small Twin Otter propeller-driven plane, she was happy to be there. I took her underground at Thompson Creek, and after experiencing some initial anxiety, she thoroughly enjoyed the tour. This was her first experience going underground.

On Saturday night, we went to the Campfire Bar in Challis where the Thompson Creek crew (geologists, miners, and drillers) and their families held a big Christmas party, complete with a lot of hooting 'n' hollering—a grand celebration indeed! Janet loved it, as did I. It gave me a chance to show her off. It was difficult to maintain our romance since I was away so much, so Janet and I took advantage of any available opportunity to spend time together. We found that we were kindred spirits; she liked travel and adventure as much as I did.

So the year ended on a high note; I had a wonderful and beautiful girlfriend and was working on an exciting project. So far, working at Thompson Creek had been the most rewarding experience of my exploration career because of the wonderful people I was working with, the leadership skills I was developing, and the geologic contributions I was making there. Can it get any better than this?

8. MARRIAGE, FATHERHOOD, AND A NEW HOME

By the end of December, we had curtailed drilling at Thompson Creek so we could evaluate the mounds of data collected so far. So in January, 1977, I was back at company headquarters in Denver, where for the next few months I devoted all my time and effort to Thompson Creek office work. A period of intense data compilation and analysis began: plotting geology and assay results on maps, drawing cross sections, and trying to construct a compelling geologic story describing the genesis of molybdenum at Thompson Creek. Spending time in the office like this was typical—and necessary—for exploration geologists. We needed to take time away from fieldwork to let the acquired data gestate in our minds. Although sitting behind a desk all day is not as fun as being in the field, it often results in exciting and unexpected discoveries. Would that be the case here, and would I have an ah-ha moment for Thompson Creek?

Yes! Once I had contoured on cross sections the molybdenum values from all the drill hole samples, I patiently looked at the distribution of molybdenum throughout the orebody; something surprising revealed itself. A portion of the mineralized zone contained substantially higher grades of molybdenum than elsewhere. We had not realized this during the period of frantic drilling. Since this higher-grade rock was in the center of the zone of mineralization, I dubbed it the high-grade core. My thought was that if mining should occur, this higher-grade zone would substantially improve the economics of the mine. The rocks had been speaking to me, and I just now heard them. What an epiphany! I was thrilled at this discovery, as were my bosses.

With renewed optimism, I flew to Chicago for a Noranda company meeting and then on to Toronto to assist Noranda's corporate mining engineers in their analysis of Thompson Creek data. By applying different mining scenarios, the engineers were trying to determine if Thompson Creek could be mined at a profit. My discovery of the high-grade core required that a different mining scenario be considered, and they set right to work on it. I also flew twice to Los Angeles to share my recent findings with Cyprus corporate geologists and engineers. At this early point in the evaluation of mining feasibility, everyone was excited and encouraged.

The next steps to be taken at Thompson Creek were in the hands of the mining engineers; on my part, I tried diligently to get approval to restart the drilling program. I thought there was more ore to be found. But my attempts were unsuccessful, and fieldwork languished. However, I did not give up, and in early June, I went to Toronto again to provide more input for the mining feasibility studies being conducted in Noranda's corporate office. The wait was extremely frustrating.

As summer came to Idaho, I returned there in early June for another season of fieldwork but not at Thompson Creek. I did, however, give a tour of Thompson Creek and an exploration update to Cyprus Mines and Noranda corporate officials, and they seemed to appreciate the implications of the high-grade zone of molybdenum mineralization. However, despite my best efforts, still no additional drilling was approved.

But there was still more exploration potential in the area around Thompson Creek, so I began a season of reconnaissance fieldwork with a new field assistant, Bill Bourcier. Bill was a lanky Oregon boy who had just graduated from Oregon State University and was spending the summer with me before heading to graduate school. About fifteen years later, I worked with Bill again, this time at Lawrence Livermore National Laboratory in Livermore, California. Bill recently recounted to me that he had wood ticks to thank for his getting the job that summer in Idaho—the person who was originally scheduled to assist me had, unfortunately, come down with Rocky Mountain spotted fever and had to go home to recuperate.

Back to that summer in Idaho, Bill and I examined the Ramey Creek molybdenum prospect a few miles west of Thompson Creek. This property was unimpressive, but the hike into it was beautiful. We saw a stately mountain goat from only about fifty feet away. Of course I had no camera with me. In addition, we examined the Little Fall Creek molybdenum prospect northeast of Sun Valley in the Pioneer Mountains. This location was not only scenic but held impressive geologic credentials as well. We noticed two mining claims had been staked there, so we would have to check out the land status later at the county courthouse before we proceeded to do more work.

Continuing our regional reconnaissance, Bill and I examined two other molybdenum properties not far away: the Little Boulder Creek prospect in the Boulder Mountains and the White Clouds deposit in the White Cloud Peaks. We thought the latter would be especially favorable country for molybdenum prospecting since ASARCO (another mining company) had recently announced the discovery of their 150-million-ton White Clouds molybdenum deposit. Bill and I took a seven-mile one-way hike to look at this deposit, which lies about eighteen miles due south of Thompson Creek. We were warmly greeted at ASARCO's field camp by a University of Idaho geology graduate student stationed there as caretaker of the camp; he was eager to share with us what he knew about the local geology. We learned later that ASARCO was not able to put this deposit into production because the area had been designated a federal wilderness area and therefore became off-limits to mining.

A little later, after examining the land-status records in the county courthouse in Hailey, Idaho, we determined that only two claims had been staked at Little Fall Creek, and the rest of the area was open for staking. So after buying some wooden posts from a local lumberyard, Bill and I returned to the Little Fall Creek prospect in early July and began staking twenty mining claims. The countryside was beautiful there in Idaho's Pioneer Mountains, but the rugged terrain was difficult to traverse while carrying heavy wooden claim posts. Measuring four inches by four inches and five feet long, we couldn't carry many posts at once. These claims supplemented the two that had been staked earlier by two individuals from Salmon, Idaho. We met them one night for dinner to discuss making a deal for their claims—they had in mind that we would give them cash right there on the spot, while we had in mind that maybe they would receive a check someday from Noranda if the property ever went into production. Despite our vastly differing expectations, Noranda subsequently leased their two claims. Interestingly, these prospectors thought they had discovered galena, a lead-sulfide mineral somewhat resembling molybdenite. However, it was, in fact, molybdenite.

After staking the claims, Bill and I geologically mapped the property and collected samples of rocks and stream sediments for shipment to a contract analytical laboratory for chemical analysis. We wanted to determine the molybdenite content of the samples and see if there were any trace elements that might indicate the presence of molybdenite at depth. What attracted us most were the outcrops of molybdenite-bearing quartz veins up to two inches wide. We knew if there were enough of these veins we could make a mine out of this property.

Our work at Little Fall Creek apparently was of interest to someone else too, because one afternoon we were buzzed three times by a low-flying helicopter. No doubt they were spies from another mining company, we thought. But we beat them to it!

One morning we stashed some cans of beer in the creek as we headed out to do fieldwork at the Little Fall Creek property. Upon returning at the end of a long day, we had instant, cool refreshment. What I wonder now is why we didn't do that more often!

Some days later, we could barely see across the property because of thick smoke coming from forest fires in California. That's a long way for smoke to travel. Not only could we smell its pungent odor, but it also caused our eyes to water.

While working at Little Fall Creek, Bill and I stayed in a motel in the mountain ski-resort town of Sun Valley, a more ritzy community than is usually associated with mining projects.

The modern-day glitz of Sun Valley contrasted sharply with the out-of-context image we saw of a Basque herder on horseback and his dog herding sheep at Little Fall Creek. This guy made a living by staying in a sheep wagon for months on end while caring for sheep roaming the hillsides. What a primitive lifestyle compared to Sun Valley, famous as America's first destination winter resort. Developed in 1936 by W. Averell Harriman to increase ridership on his Union Pacific Railroad passenger trains, Sun Valley was modeled after similar destination mountain resorts in the European Alps.

Sun Valley is where Ernest Hemingway lived off and on from 1939 to 1958 while writing *For Whom the Bell Tolls* and other novels. He later moved south of Sun Valley to the Wood River Valley town of Ketchum, Idaho. He was buried there after taking his life in 1961.

When we finished mapping at Little Fall Creek, I was given the go-ahead to do detailed surface geologic mapping at Thompson Creek. At least that showed some continuing interest in the property on Noranda management's part. Since it was already late September and winter weather would be threatening soon, I immediately went right back there. Bill Bourcier had to leave to go to graduate school by this time, so Asa Clark,

A cool, refreshing can of beer put in the creek
for our day's end refreshment, 1977.

one of my field assistants from the previous summer, rejoined me to assist. The geology we were mapping continued to intrigue us, and the weather cooperated pretty well with clear, crisp days and cold nights, although on some days we did have to withstand snow and pretty cold temperatures. The deer liked this weather also; we saw many of them while mapping.

Asa and I plotted on our map the fine details of rock types and geologic structures we encountered, plus the location of any quartz-molybdenite veins. Most rock exposures consisted of post-ore volcanic rocks—that is, rocks that had been deposited later on top of the ore-bearing Thompson Creek stock. These post-ore volcanic rocks just got in our way; in only a few places did the Thompson Creek stock and its molybdenite poke through the overlying cover. We had to do a lot of guessing as to what lay below those volcanic rocks. We completed the mapping before snow covered everything and then began waiting again for the go-ahead to restart the drilling.

Between mapping stints at Little Fall Creek and Thompson Creek, Janet and I found time to be married in Evergreen, Colorado, on September 10, 1977. Included in the deal was Janet's cute little eleven-year-old daughter, Debra; the two of us got along famously.

In December, I was transferred to Noranda's Missoula, Montana, office, which was responsible for exploration of the Great Northwest. So Janet, Deb, and I moved there, where we had three and a half years of further adventure. As William Feather, a noted American publisher in the early 1900s, said, "One way to get the most out of life is to look upon it as an adventure," and that is what we were doing. Janet and I like to joke that because I was marrying the boss' secretary, he had banished us to Montana! Not true—the banishing part, that is!

Janet and I immediately discovered what a beautiful spot we had been moved to. Missoula sits in the broad Bitterroot Valley of mountainous western Montana where the Bitterroot, Clark Fork, and Blackfoot rivers converge. Janet, Deb, and I once collected delicious chokecherries along the banks of the

Blackfoot; they make scrumptious jelly. From our front yard on the slopes above downtown Missoula, we enjoyed great views to the west of the Bitterroot Range, with its beautiful red sunsets. Contrary to what most people think about Montana, the winters in Missoula were not too cold or snowy. However, frigid Butte, Montana, was not far away. Missoula is home to the University of Montana, and on the slopes of Mt. Sentinel, high above the campus, we could see the iconic M representing the town and the university.

But what are those horizontal lines clearly visible on the slopes of Mt. Sentinel? It turns out they are ice age glacial features—horizontal terraces called strandlines that were created by varying water levels of Glacial Lake Missoula. You could think of these features as ancient bathtub rings. Lake Missoula was a huge glacial lake created about twelve thousand years ago during the Pleistocene epoch (11,700–2.6 million years ago) when a continental glacier moved south from Canada. Its ice blocked the regional drainage system and impounded glacial meltwater up to two thousand feet deep over much of western Montana. This lake was huge; it contained more water than the combined volume of Lakes Erie and Ontario.

Glacial lakes were common in the northern parts of the United States during the ice age. Evidence and remnants of other glacial lakes exist, such as the Great Salt Lake (Glacial Lake Bonneville); Pyramid Lake, Nevada (Glacial Lake Lahontan); and Glacial Lake Hitchcock on the Connecticut River in New England. I had studied the varves of glacial Lake Hitchcock while on a geology field trip at Dartmouth a few years earlier. A varve is a thin layer of clayey sediment that is deposited at the bottom of glacial lakes. One varve represents one year of deposition, so by counting the varves one can determine how long the lake existed—about four thousand years in the case of Lake Hitchcock. Interestingly enough, my great-grandfather Eustache Lamarre moved from Quebec to New Hampshire, where, in 1896, he started the Lamarre Brickyard in Bath, my hometown. And what was the source of clay from which the bricks were made?—the clayey sediments deposited 15,000–12,000 years ago in nearby Glacial Lake Hitchcock! There was geology at play in my lineage.

As I got settled into Missoula, still in the forefront of my mind was, *What would become of Thompson Creek?*

9. A LOVE LOST

Moving to Montana was a job transfer and a promotion, and it proved to be a positive one. I had been assigned to Noranda's Northwest district office in Missoula to initiate a molybdenum exploration program in the northern Rocky Mountains of Montana, Idaho, and Washington. This was exciting—new country and geology to see, along with the opportunity to develop and lead a new broad-based exploration program. Already working out of that office were fellow Noranda geologists Vic Chevillon, scouring the same countryside for gold, and Hart Baitis, exploring for uranium. At Cobalt, Idaho, Greg Hahn was leading Noranda's cobalt exploration program at the old Blackbird mine west of Salmon. Bill Schwerin, our invaluable technician, provided field assistance to all of us geologists. At that time, Noranda had a large presence in the Northwest with Gordon Hughes serving as the district manager.

As is typical for an exploration geologist (at least in the more northerly latitudes), January was a data evaluation and report-writing month since the weather prevented us from going into the field. I started drafting maps and writing up the results of our 1977 exploration work at Thompson Creek and Little Fall Creek. Unfortunately, and to my utter shock and surprise, the Thompson Creek project report turned into a final report because, at the end of January, Noranda headquarters informed us they had not been able to negotiate a satisfactory joint-venture mining agreement with Cyprus Mines. Noranda's involvement at Thompson Creek ended, even though I had increased the ore reserves there by 28 percent and identified a forty-five-million-ton, high-grade core. What a crushing, personal disappointment Noranda's decision was. I would have loved to have stayed on as mine geologist at Thompson Creek, but that was not to be. I had felt so comfortable at Thompson Creek that working there had given me a sense of place in the world. I would now have to find a new place.

Thompson Creek did become a mine but without Noranda's participation. In 1983, Cyprus Mines created the subsidiary Thompson Creek Metals Company to mine the Thompson Creek ore by open-pit methods. By 2010, the mine was the fourth largest producer of molybdenum in the world, employing about eight hundred workers to mine and process the two hundred million tons of ore. That number of workers was a large increase in the number of jobs available in this rural, primarily ranching area and a boon to the local economy and the state. I like to think that I contributed to this mining operation by identifying the deposit's high-grade core; I suspect that my discovery greatly enhanced the economic credentials of the mine. By mining the higher-grade ore first, the initial capital investment could be recovered faster.

Some years later, my wife, Janet, and I toured the open-pit mine in the company of the Thompson Creek mine geologist. We drove right through the heart of the orebody. How excited I was to see the ore-bearing Thompson Creek stock fully exposed where the overlying Challis Volcanics had been removed. When open-pit mining began, about two years of heavy-equipment work were required to strip away this overburden (the post-ore volcanic rocks) to reach the orebody. I don't think the mine geologist appreciated how thrilled I was to be there that day.

By February, the weather had cleared enough that I was able to go to northeastern Washington to examine two molybdenum properties: the Western Molybdenum Company property near Chewelah, and the Vanasse molybdenum property near the town of Kettle Falls on the Columbia River, a property that Noranda had leased the year before. I enjoyed getting back into the field, even though I had to dig through snow to see the outcrops.

At Vanasse the previous summer, Noranda geologists had drilled four shallow exploratory drill holes that had intersected molybdenite in sparse to modest amounts. To further our understanding of the geologic system that had produced the molybdenite, I spent a few days geologically mapping in detail one and a half square miles of this stockwork quartz-molybdenite vein system. I was snowed out a few days, and on another much warmer day, I came face-to-face with a rattlesnake as I was climbing up a cliff face. Apparently I had interrupted the snake as it was trying to warm itself on a ledge. My mapping showed promise for the property, so later in the year I developed a work plan and supervised the drilling of two exploratory drill holes for a total of 1,394 feet. We drilled into the most highly altered and mineralized part of the Vanasse quartz monzonite, but unfortunately the results were unspectacular, and we dropped the property a few months later.

While in Washington that spring, I looked at other properties near Republic, Colville, and Tonasket in the northern part of the state near the Colville Indian Reservation. I don't remember much about these properties, but a notation about one of them in my field notebook reads, "It's a dog!" Regrettably, I never did get to work farther west in the Cascade Range; that would have been a scenic thrill among the volcanoes there. Janet joined me on one of my trips to Washington, and we made a tourist visit to Trail, British Columbia, just over the Canadian border. Perched on the mighty Columbia River, the town of Trail was home to a lead/zinc smelter.

During this timeframe, I made my first foray into the political arena. Along with Vic Chevillon and Gordon Hughes, I flew from Missoula to Spokane, Washington, where the three of us testified at a US House of Representatives public hearing on mining law reform. Environmental interests were strong at the time and were putting increasing pressure on federal lawmakers to limit access to public land for exploration and mining. Naturally, this was counter to our interests and not good for the country, we believed. Fortunately, no changes to the law were made.

Of course I was eager to learn about the geology of my new surroundings in Montana, so one weekend we took a family field trip to Glacier National Park, a few hours' drive north of Missoula near the Canadian border. As we drove along the twisty and narrow Going-to-the-Sun Road in the heart of the park, we were greeted with spectacular mountain scenery that prompted me to comment, "My stock of landscape adjectives is running low." We had a treat when we stopped to watch a large moose having lunch in one of the marshes. I also remember seeing textbook examples of stromatolites—mounds that resemble large cabbage heads that were built by ancient colonies of primitive blue-green algae—in one of the road cuts in the Precambrian Belt Series rocks. Of course, collecting rocks is not permitted in national parks.

Glacier National Park is a natural outdoor geology museum where we saw beautifully preserved sedimentary structures in the rocks, including mud cracks, raindrop impressions, and ripple marks. The first feature told us these rocks were once layers of mud that cracked in the baking sun on a mudflat; the raindrop impressions told us the mud was then pelted by a passing rainstorm. All this happened about 1.4 billion years ago during Precambrian time when sediments piled up in a huge basin forming layers totaling ten miles thick. The rocks, indeed, were speaking to me, even on non-workdays.

These Precambrian Belt Series rocks are pervasive throughout northwestern Montana and northern Idaho and consist of metamorphosed sandstones and mudstones. I would get to see many of these rocks in the next few years. Even though it is probably obvious to the casual observer that the rugged mountains and valleys of Glacier National Park were carved by glaciers, it is not at all obvious that the park sits on a giant slab of Precambrian rocks that overlies much *younger* sedimentary rocks of the High Plains. The Precambrian rocks had slid eastward about fifty miles on a nearly flat surface called the Lewis Overthrust fault. How such movement could occur is still being debated among geologists. Contrary to a cardinal rule of sedimentary geology, older rocks here lie on top of younger rocks. I regret not having visited this supreme example of Precambrian geology and glaciated mountain scenery more often while we were living so close by. Perhaps another trip to Glacier should be on my bucket list.

In May 1978, I flew east to Middletown, Ohio, where Geoff Snow and I talked with Armco Steel Corporation executives about their joining Noranda in our molybdenum exploration program. Armco was interested because steel makers use molybdenum to make high-strength alloy steel, and Armco needed a lot of it. After some negotiations, Armco decided to enter a joint venture with us by providing a share of the exploration funding required. They clearly envisioned the possibility of owning a share of a molybdenum mine, and we enjoyed a productive relationship with them for the next three years. Our objective that first year was to "discover at least one drillable molybdenum target by December 31, 1978, at a cost not to exceed $150,000." By the end of the year, we had accomplished that objective.

During the winter and spring of 1978, I had been conducting a literature survey of the geology of western Montana and had identified areas that might be favorable for molybdenum deposits. The literature search required spending time at the University of Montana library in Missoula, looking primarily at technical reports published by the US Geological Survey, the Montana Bureau of Mines and Geology, and the University of Montana. From this research, I then created a work plan for the summer's field season, complete with goals of the project, personnel requirements, and a budget.

To implement the plan, I interviewed and hired six summer field geologists to assist me, working in two-person teams, each in a different part of the state. You don't need to remember their names, but I list them because they were special to me: Dave Hyde, Bill Murphy, Steve Richardson, Alan Sweide, Bruce Nickelsen, and Bob Thompson. These guys were either recent geology graduates or geology students eager to gain relevant experience in a summer job. I made them responsible for conducting large-scale reconnaissance geologic mapping, rock and stream-sediment sampling for geochemical analysis, prospect evaluation, detailed mapping of prospects that looked attractive, and in some cases claim staking.

Since these field assistants had little if any previous field experience, I viewed my role as one of teacher as much as boss. I enjoyed the role of guiding these young people in understanding that the rocks would speak to them if given a chance. I visited each crew in the field on a rotating basis, reviewed what they had seen in the field on their traverses, and guided them in what steps to take next. I also kept in close contact by telephone. I thoroughly enjoyed their eagerness and enthusiasm for the work they were doing. These students reminded me of my earlier days in exploration, and I hoped I could mentor them as effectively as I had been mentored by Geoff Snow and Bob Hodder.

Together that summer, we conducted exploration in the Rocky Mountains of Montana from near Glacier National Park in the north to near Yellowstone National Park in the south. This project took us to places most people have probably never heard of: Flint Creek Range, Castle Mountains, Little Belt Mountains, Big Belt Mountains, Tobacco Root Mountains, Bear Paw Mountains, Crazy Mountains, and the Absaroka Range. Most of these mountain ranges were glaciated during the last ice age, and we were often greeted by spectacular U-shaped valleys and sharp peaks, both characteristic glacial features.

In a short time, we succeeded in identifying attractive molybdenum prospects in the Flint Creek Range just west of the town of Deer Lodge, Montana. There, we staked fifty-five mining claims on East Goat Mountain and thirty-five claims on Mt. Rose, and were only about five days too late to stake claims on another target; a competitor company had beaten us to the punch. One had to be aggressive because there was a lot of exploration activity in Montana being conducted by major and not so major mining companies, such as ASARCO, Bear Creek Mining Co., Utah International, Homestake Mining Co., Amax, and Phelps Dodge. By the 1980s, most of the major oil companies had created metals exploration programs also.

What attracted us to East Goat Mountain was the presence of good-looking quartz-molybdenite veins in a stockwork-vein system exposed in outcrops of intrusive rocks on the mountain's steep western slope. Some samples of vein material contained up to 0.2 percent molybdenum. While claims were being staked for us by a contractor, two of my field assistants and I explored some old adits and mapped the surface geology in detail. We were impressed with what we saw—especially the quartz-molybdenite veins up to eight feet thick exposed in an adit—even though we sometimes had to walk in snow up to four feet deep to reach the outcrops. One day as we were flying into the property in a helicopter from Deer Lodge, we saw a herd of about fifty elk grazing in a meadow below us. What an impressive sight these beautiful animals provided. The Flint Creek Range was a thrilling place to work, not only for its geology and wildlife scenes but also for its sights of glaciated high mountain peaks reaching elevations of ten thousand feet.

Noranda's office in Missoula had a Honda 90 trail bike, so one day I decided to use it at East Goat Mountain, thinking it would be easier to ride the trail bike into the property than to hike. Big mistake. After bouncing along atop rocks that littered our so-called hiking trail, I came to an unanticipated dead stop. It turns out one of the many boulders I hit had sheared off the oil-pan drain plug, and I had lost all the engine oil. Hiking was a much better way to go—and now the only way! On the return to my truck, I had to push the motorcycle, grumbling all the way.

While prospecting that summer in the Pioneer Mountains of southwestern Montana around the town of Melrose, we again experienced steep, mountainous terrain, much of it above tree line at elevations up to ten thousand feet. The Pioneer Mountains rim the Big Hole Basin, a gorgeous and peaceful spot with the Big Hole River lazily flowing through hay fields and green meadows. I saw many of the iconic beaverslide hay stackers that were invented here in 1910 to help put the dried hay into twenty-ton stacks. By way of humorous exaggeration, the Big Hole Basin was once called "The Valley of 10,000 Haystacks." This site is suitable for the cover of *Sunset Magazine*, and fly fishermen love the area. Someone once described the Big Hole Basin as "the front parlor of Heaven," and I'm inclined to agree.

For our work in the Pioneers, we contracted with Hawkins & Powers Aviation out of Greybull, Wyoming, to provide helicopter support—in a Jet Ranger helicopter we were able to cover a lot of territory quickly. In all, we evaluated sixty square miles in the northern Pioneer Range. As in Alaska a few summers earlier, the pilot dropped us off at a convenient spot in the morning, we did geologic reconnaissance all day, and then the pilot returned to a designated spot at the end of the day to pick us up. While flying, it was fun to catch the updrafts of the air currents and ride the thermals like raptors. One day we discovered old, twenty-foot-tall beehive-shaped brick kilns sitting all alone among cottonwood trees in a pretty, hidden valley. Miners in the late 1800s had built them to make charcoal from local trees, and the charcoal was then used to smelt ore. Even though this area was the site of much past and present mineral industry activity, we found no molybdenum prospects worthy of pursuit. We did, however, discover two stately mountain goats watching us closely from a hillside one day.

This part of Montana, being Lewis and Clark country and Indian country, offered intriguing place names that I recognized from history books: Chief Joseph of the Nez Perce Indians;

The author commuting to work in the
Pioneer Mountains of Montana, 1978.

Blackfoot; and Flathead. Montana even has a Lewis and Clark County and a Lewis and Clark National Forest. Interestingly enough, daughter Debra attended Lewis and Clark Elementary School in Missoula, and daughter Meaghan graduated from Lewis and Clark College in Portland, Oregon.

While sitting atop the Castle Mountains late one summer afternoon with eager field assistant Steve Richardson, we watched huge cumulus clouds rise like fury in the late afternoon heat, as if they were boiling. "That could generate a tornado," Steve commented. A few days later while hiking up one of the mountain valleys there, we were surprised to see that all the trees on one side of the valley had been knocked down in one direction, and those on the other side were lying in the opposite direction. We surmised that a tornado must have passed through here in the not too distant past but not that week. Even though this happened more than thirty years ago as I write this, it stands out to me as clearly as if it were yesterday.

On one of our prospecting trips to central Montana, two of my assistants and I drove along the Missouri River, through the city of Great Falls and then on to a collection of a few buildings beside the road called Monarch. Monarch was in competition with nearby Neihart for being the smallest town around. From there we examined exposures of molybdenite at the old Big Ben mine in the Little Belt Mountains, the setting for these two towns. Even though it was summer—August 17—snow still remained in the high country. A little farther on, we passed through the town of Ringling, Montana, named after the Ringling brothers of Ringling Brothers Circus fame. At one time the Ringlings owned ranch land in the area amounting to about one hundred thousand acres; they even considered establishing their circus headquarters there.

All this prospecting in Montana led to cherished kudos from my boss, Geoff Snow. After examining one prospect, I wrote a report of our property evaluation, and Geoff sent me the following note: "It is truly a pleasure to get a piece of work like this report. Not only do I know what and why you recommend, you gave me sufficient data that I may reach my own conclusions. As it is I concur with yours. Thanks and keep up the good work."

Later that summer, our work took us to Philipsburg, Montana, where, even though the rocks were not too exciting, the wildlife was; seeing three black bears sauntering along the hillside one day was a thrill. We gave them wide berth. Philipsburg is near Butte, Montana, home of the huge Berkeley Pit, the largest copper mine in Montana at the time. The mine straddles the Continental Divide where copper had been discovered in the Boulder Batholith, a late-Cretaceous (73–78 million years old) mass of granite that covers thousands of square miles of mountainous terrain around and between Butte and Helena. The batholith received its name from the numerous boulders that result from weathering of the granite.

Butte's Berkeley Pit began as a series of underground mines in the late 1800s, as so many mines did. By the 1890s, it produced 40 percent of the world's supply of copper. In 1955, the underground mines were converted to a lower-cost open-pit mine to recover the lower-grade ore. Anaconda Copper Company employed nine thousand miners here in its heyday. Mining ceased in the huge Berkeley Pit in 1982, and it is now partly filled with water, but small-scale mining continues nearby. The copper ore was so rich that the citizens of Butte referred to the area as "the richest hill on earth."

In the early days of mining there, Butte was a rough-and-tumble copper boomtown with a notorious red-light district, hundreds of saloons and gambling houses, and ruthlessness that bordered on lawlessness. It seems that most of the western mining towns of the late 1800s and early 1900s, such as Jerome, Arizona, and Virginia City, Nevada, were in competition for the title of "the baddest town in the west." It is said that in Pioche, Nevada, seventy-two people were killed by gunshots before the first person died of natural causes. Pioche was an important silver mining town and was also the most notorious mining camp in Nevada.

While working around Butte, our prospecting took us out one night to look for exposures of scheelite. This tungsten mineral sometimes accompanies molybdenite, and, using a black light, it can easily be identified at night because of its fluorescent properties. What fun! One Sunday I took some time off and went panning for gold and sapphires in a nearby creek. However, I didn't find enough to permit me to retire.

East of Butte at Three Forks, Montana, are the headwaters of the Missouri River. Here, the Jefferson, Gallatin, and Madison Rivers merge to create the Missouri, that mighty river draining the east side of the Northern Rockies. This storied river carries sand and silt from near Butte north and eastward across Montana, through the Great Plains' wheat fields of North and South Dakota, past Council Bluffs, Iowa, then to the Mississippi River where it eventually is deposited as the vast delta at New Orleans. Because of the conveyor belt action of the Missouri River, the Mississippi River delta is comprised in large part of detritus from the Rocky Mountains. Terrain that is being worn down in one place (Montana) is being built up in another (Louisiana). Without this process, there would have been no land on which to build the city of New Orleans. Geology in action again.

Ironically, because levees and dams have been built along the course of the Mississippi River for flood protection, sand and silt are no longer being deposited in the immediate vicinity of New Orleans. Since deltas naturally compact and sink over time as the weight of overlying sediments expels water from them, the city of New Orleans is sinking as well. That city now lies about four feet below sea level, and many of its houses are built on stilts. It's not nice to fool with Mother Nature, even if you are the US Army Corps of Engineers.

In late September, all Noranda Exploration, Inc., geologists convened in Spokane, Washington, for the first annual meeting of the company's Witchers & Dowsers Society. This society, the brainchild of Geoff Snow, went on to conduct a field trip each year to visit and learn about different types of ore deposits. Again, this was a result of Geoff's insistence that company geologists continuously expand and broaden their geologic expertise. The theme this first year of the society's existence was Precambrian Belt Series rocks of Idaho and Montana and their contained ore deposits. These rocks are rift-deposited (remember, a rift is a fault-bounded, linear valley) metasedimentary rocks that cover much of northern Idaho and western Montana and are hosts for deposits of copper-silver-lead-zinc-cobalt in the Coeur d'Alene Mining District of northern Idaho.

Traveling by chartered bus and led by resident expert Don Winston, a geology professor from the University of Montana, we looked at outcrops around Idaho's Pend Oreille Lake on the way to ASARCO's Coeur Mine and Hecla Mining Company's Lucky Friday mine. Both were underground operations, and the tours we took were fascinating—I always loved going underground, but you already know that. What a thrill to walk onto a skip (an elevator) with eight to ten other geologists and be lowered thousands of feet into the bowels of the earth. These mines are in the Silver Belt of northern Idaho near the mining camp towns of Kellogg and Wallace.

As I had seen at Sudbury, Ontario, while in graduate school, the hills around these towns had been defoliated by sulfurous gases emitted from the smelter stacks. In spite of this, this part of Idaho is famous for attracting outdoorsmen who love hunting and fishing.

Just as Arizona is the country's largest copper-producing state, northern Idaho is the largest silver-producing area in the country. From 1884 to 1984, the Silver Belt produced just under a billion ounces of silver. Other major metal-producing states are Colorado, which produces the most molybdenum, and Nevada, which mines the most gold.

University of Montana geology Professor Don Winston instructs Noranda geologists at a Witchers & Dowsers Society field trip in western Montana, 1978.

After leaving the Silver Belt, we examined outcrops as we drove to Missoula, then to Melrose, Montana, with a final destination at Noranda's cobalt exploration property at Cobalt, Idaho. The element cobalt was an attractive exploration target because of its use in cobalt-samarium magnets that can be manufactured in small sizes for use in electronic equipment. During our drive there, we had some excitement when our large tour bus got hung up on a tight corner of the dirt road leading to the mine. We, the passengers, had to help push and dig the bus out. Tour buses don't usually travel that road.

There were about thirty male geologists on this field trip, and they represented a typical cross section of exploration geologists in the industry at the time. White males constituted the vast majority of industry geologists, and many of them—if not most—sported beards. Female geologists were becoming more prevalent in the industry as time went on, but until the early 1980s, most women were not interested in becoming field geologists. None worked for Noranda until 1978.

There was precious little racial diversity among exploration geologists then either; virtually all were Caucasian. Education-wise, among us geologists the most common college degree was a master's. A few had PhD degrees or had stopped at a bachelor's. A PhD certainly was not required, however, the general thinking was that more education beyond a bachelor's degree was helpful. The most prevalent educational providers of exploration geologists at Noranda were Dartmouth College and the Colorado School of Mines. As a Dartmouth graduate, Geoff's college ties were strong, and he liked to hire fellow alums. After Dartmouth and the Colorado School of Mines, the University of Arizona, Oregon State University, the University of Oregon, and the University of California campuses were well represented.

The field trip ended back in Missoula, and all agreed it had been time well spent. We were all much more knowledgeable now about the ore deposits of the Belt Series rocks.

Later that year, I examined a silver prospect near Flathead Lake in northern Montana. This property belonged to Congdon and Carey, Ltd., the Denver-based natural resources company that owned the Gem Park property in Colorado that I had drilled in 1975. My old friend Dolf Fieldman was still their geologist. I don't remember much about the Flathead Lake prospect itself, but I do remember that it was in a beautiful location southwest of Glacier National Park in the agriculturally productive Flathead Valley. Nature plays around with four colors here: yellow of the grass, silver and blue of the lake, and dark green of the fir trees on the hillsides. A springtime visit to see the blossoms on the cherry and apple trees around the lake was always fun. This valley is the southern extension of the Rocky Mountain Trench that extends well into Canada. During the last ice age, a glacier filled this trench to a short distance south of Flathead Lake. Debris left by the melting glacial ice as it retreated filled the valley floor and is responsible for its fertility.

In August, I was excited to begin another major molybdenum exploration project, called Liver Peak. I started geologically mapping and sampling this molybdenum property in the Bitterroot Range near Thompson Falls, Montana, where Noranda held fifty-eight mining claims. Discovered by others in 1968 by sampling of silt in streambeds, the property had been explored by two other companies before Noranda acquired it in 1974 from Bear Creek Mining Company. In 1977, other Noranda geologists had drilled three exploratory boreholes for a total of 9,612 feet and found low-grade molybdenum in quartz veins.

It was now the end of the field season, so my assistants had gone home, and I was left to evaluate the results of that 1977 drilling at Liver Peak. It looked to me like there was good potential to discover ore-grade molybdenum deeper in the metasedimentary rocks of the hydrothermally altered Belt Series, so I developed an exploration program. By the end of the year, I completed field mapping of the property and had evaluated all the geologic data available. The exploration potential still looked good, and I was eager to initiate a deep-drilling program the following summer as soon as the snow melted. *Would this turn out to be as exciting and rewarding as Thompson Creek?* I wondered.

10. THE ELUSIVE INTRUSIVE

Thanks to our success last year in identifying attractive prospects in Montana, in 1979 I expanded the molybdenum reconnaissance program into Idaho; favorable geology did not stop at the state line, after all. However, I had to wait for the winter weather to clear before I could get into the field, so I sat at my desk in Noranda's Missoula office and poured over the data we had collected from the Liver Peak and East Goat Mountain projects. I needed to determine where the best drill targets were since I wanted to start drilling there as soon as road access was available in late spring. Because the US Forest Service had to sign off on our exploration plans before drilling could begin, I also hurriedly developed drilling plans for both projects. Time was of the essence.

However, a bad case of cabin fever soon set in. As a remedy, one day in February I drove to Challis, Idaho, to look at Cyprus Mines Corporation's drilling operations at Thompson Creek where they were continuing (without us) their successful ore delineation. How bittersweet it was to see work progressing there without me. At least I could continue my friendship with Floyd Marshall and his wife; I spent an enjoyable night with them in Challis.

That winter I also participated in the political arena again. After flying to Washington, DC, I spent four days lobbying Congress about a new mining bill under consideration. I met with staff members from the offices of Montana Senators Baucus and Melcher and Montana Congressman Marlenee, as well as with representatives of the New Hampshire and New York Congressional delegations. Being pretty naïve and not particularly politically savvy, I thought the experience would be a good civics lesson. I'd like to think I made a difference because public lands remained accessible to exploration and mining. My main argument was that the country's well-being and future depended on the wise use of all our natural resources, whether they were contained in fields, forests, waters, or the Earth. Since mineral resources are scarce, we needed to be able to mine them wherever they existed.

When the winter weather eased a little, I made a long drive to southwestern Idaho to begin the Idaho molybdenum reconnaissance program. Accompanying me were Dave Hyde and Rick Bellows, the first of my summer field geologists to arrive that year. Dave must have had a good time working for me the previous summer because he decided to come back for another tour of duty. Our objective along the Payette River north of Boise was to examine prospects in the Idaho Porphyry Belt where outcrops of the Idaho Batholith were cut by dikes (tabular intrusions of igneous rock that cross through older rocks) and quartz veins. Drill core from other companies' previous exploration of one property, the Morning Mist prospect, was stored at the University of Idaho campus, so, after examining the property in the field, Dave, Rick, and I drove to the university town of Moscow in northern Idaho—more long-distance driving. We wanted to examine the core to make a further assessment of the property's merit, which we did. Unfortunately, we were disappointed to learn that the core did not have geologic features suggestive of an orebody.

Nevertheless, we took advantage of the trip to learn about the local geology. Consisting of rolling hills of loess (rhymes with bus), which is wind-blown glacial silt, the area around Moscow in Idaho and adjacent Washington is called the Palouse Hills. The word "loess" is the German word for "loose," referring to the fact that the soils are unconsolidated. The loess of these hills was derived from glacial outwash eroded out of the Cascade Range and blown eastward by ice age winds. Eastern Washington contains the thickest loess deposits in the United States, up to three hundred feet deep in some places. These hills are extremely fertile and productive; acres and acres of wheat and legumes flourish here, enabling more wheat per acre to be harvested than in any other place in the country. Remember field assistant Asa Clark from last year? This is where he was raised.

Fortunately for the farmers, this area was spared the destruction wrought by flood waters coming from Glacial Lake Missoula toward the end of the ice age. At that time, the ice dam that created the glacial lake gave way and allowed billions of gallons of water to cascade across northern Idaho, Washington, and Oregon, scouring canyons along the way. This was the largest documented flood in geologic history. All of this geology has nothing to do with ore deposits, but it does have a lot to do with the economic prosperity of the area.

Once back home in Missoula, I left again almost immediately to begin the summer's exploration work in Montana. Along with three additional new field assistants that I had hired (Bob Kemp and husband-and-wife team George and Cay Kendrick), I went to the area around Lewistown—almost the geographic center of the state—where we looked at several properties in the Judith Mountains. Over a three-week period, we mapped and sampled eleven areas of potentially favorable geology and examined some old underground workings in what geologists call the Central Montana Alkalic Province. None of these prospects was worth pursuing further, but the sight of the wide-open countryside of the northern Great Plains' vanishing prairie off to the east was well worth the visit. Had our vision been better, we could have seen Minneapolis from the summit of the Judith Mountains; there was nothing to impede our eastward view for 750 miles, except for the curvature of the Earth.

We have Thomas Jefferson to thank for this being part of the United States, even though his Louisiana Purchase was not popular back in 1803. Had we been there at that time, we would have seen vast expanses of semiarid, native-grass-covered plains adorned with thousands of grazing buffalo. No buffalo remain, thanks to the buffalo hunters, and there isn't much intact native grassland left either, thanks to the success of the homesteaders' plows.

Fascinating to me *now* is the fact that this Central Montana Alkalic Province contains kimberlite pipes (vertical, cone-shaped bodies of dark-colored igneous rock) that sometimes contain diamonds. There are no diamond mines there, but one pipe does contain microscopic diamonds. When I worked there in 1979, the kimberlite pipes had not yet been discovered, and no one realized that the geology was favorable for diamonds. This demonstrates how exploration and geologic knowledge evolve over time, requiring continuous learning on the part of geologists.

By May we had a drilling company under contract and all federal permits in place to begin drilling at Liver Peak. After plowing snow from two miles of access road, we began drilling on May 15, 1979. Our objective this year was to drill a five-thousand-foot hole into the center of the hydrothermal system in an attempt to locate the intrusive rock that we deduced to be the source of the alteration and mineralization visible in the Belt Series rocks at the surface and in drill core. Our deep drilling would test this conceptual geologic model. I hired two additional field assistants, Randy Moore and Larry Campbell, to work on this project. One of the first things I showed them was how to geologically log drill core, a key and necessary skill. The two of them caught on quickly.

After ten years of exploration by various companies at Liver Peak, in August we intersected the elusive buried intrusion at a depth of 2,534 feet in drill hole LP-4. Hurray! What good news! Heretofore, no igneous rocks were known to be present at the property; however, we knew they had to be there as the source of molybdenite found in quartz veins. We named the intrusive rock the Liver Peak Intrusive Complex because it consists of five distinct magmatic phases. We bottomed the hole (stopped drilling) at 4,788 feet in late November because of "tight hole" conditions—that is, we were twisting the drill rods off in the hole. Not a good thing. And the weather was getting really bad as well. Old Man Winter was on his way.

However, there was additional good news. In that deep drill hole, we also intersected 970 feet of altered Belt Series rock averaging 0.12 percent molybdenite. When combined with comparable intercepts in nearby drill holes, we estimated the potential for as much as four hundred million tons of rock averaging 0.14 percent molybdenite. We called this zone the Upper "Ore" Zone. The word "ore" is in quotation marks because this molybdenite grade and tonnage were not yet sufficient to meet the definition of ore—rock that can be mined at a profit. However, we were optimistic. Though not as high grade as we would have liked, sufficiently large tonnage could constitute an economically viable deposit, depending upon the molybdenum price, of course.

But there was more! Near the bottom of this drill hole, we intersected a second and heretofore unknown zone of mineralization, this time in the intrusive rock: 620 feet of drill core contained 0.13 percent molybdenite in quartz-molybdenite stockwork veins. In some intervals, I counted twenty veins per foot in the drill core, resembling the stockwork ore being mined at the Climax and Henderson molybdenum mines in Colorado. We named this the Lower "Ore" Zone. Original, huh? In a previously drilled hole farther north, we labeled a zone of mineralization the North "Ore" Zone. Our findings were extremely encouraging and exciting because the rock type associations, alteration characteristics, type and age of mineralization (38.6 million years) all indicated the geology was favorable for hosting a molybdenum ore deposit. By the way, we knew the precise age of the mineralization because the US Geological Survey had radiometrically dated some of the alteration minerals for us. Thank you, USGS!

That same summer while drilling was underway, we conducted detailed surface geologic mapping of the property. To our surprise, we discovered that hidden away in some brush were outcrops of the intrusive rock we had intersected in drill core. After all this time, the elusive intrusion had been found at the surface!

Also that summer we staked 140 additional mining claims around the existing claim block at Liver Peak and acquired a lease on fourteen others. These additional claims provided full coverage of the main area of mineralization—insurance, if you will—and provided access to a possible adit site from the Thompson River Valley. Should underground mining be warranted, we could drive (dig) an adit in from the valley floor to mine the deposit from below by block-caving methods. These actions clearly demonstrated the optimism we were all experiencing at the time. For the two, young field geologists Randy and Larry, this was an especially exciting time—they were drilling "ore."

Thompson Falls is the town nearest to Liver Peak, only five miles away, and I spent many nights there. This Lolo National Forest location is about a two-hour drive from Missoula and pretty close to Paradise—the town of Paradise, Montana, that is. Since we now had an ongoing presence in Thompson Falls, we rented a house there as accommodations for Randy and Larry. That was much more comfortable than a motel room and also served as a suitable field office. On one of my many driving trips to Thompson Falls to review progress, I saw in a meadow right beside the road four herds of about forty-five bighorn sheep each. What a sight! Thirteen-year-old Deb accompanied me on some of these trips to Liver Peak; too bad she wasn't with me that time.

While drilling was underway at Liver Peak, Bob Kemp (another of my field assistants) and the Kendricks—my northern Montana reconnaissance crew—were prospecting northwest of Helena, Montana. They evaluated the molybdenum potential of about one thousand square miles of real estate in the Lewis and Clark National Forest. I had sent them to this area because of the recent discovery by a competitor company of a small stockwork molybdenum deposit at Bald Butte. My field crew was successful in discovering three prospects that warranted more detailed attention later.

It turns out that some of this work was near the town of Lincoln. Unbeknownst to us at the time, this is where Ted Kaczynski, the Unabomber, was fabricating mail bombs in his remote cabin. It's a good thing we didn't run into him!

Prospecting also took me to Montana's Absaroka Range just north of Yellowstone National Park. As you can imagine, the scenery was gorgeous from atop 10,921-foot Emigrant Peak. I had a top-of-the-world view from there. The geology was stimulating also, because from this vantage point I looked south to the surface expression of the Yellowstone supervolcano that underlies the park.

Yellowstone last erupted 640,000 years ago, producing a caldera forty-five miles across. As you may remember, this caldera is the youngest and easternmost of a string of ten calderas, most of which are buried beneath the Snake River Plain in adjacent Idaho. To be precise, three calderas coalesce at Yellowstone—and thus the name supervolcano—and they were created by eruptions of titanic scale. Ash from these eruptions was carried as far away as present-day Iowa and Louisiana. One eruption, the gigantic Huckleberry Ridge eruption 2.1 million years ago, was one of the largest volcanic blasts in the history of the Earth. Imagine it, belching six thousand times more volcanic ejecta into the air than did the Mount St. Helens eruption of 1980. As I collected samples of Mount St. Helens' gray volcanic ash from my driveway in Missoula the next year, I thought *that* eruption was big enough. Yellowstone has erupted three times in the past two million years, making it perhaps the scariest of the world's supervolcanoes. Another eruption will occur one day from Yellowstone, and when that happens, it would be wise not to be too close.

What we see today at Yellowstone National Park is a huge hydrothermal (hot water) system of geysers, fumaroles (steam vents), mud pots, and mud volcanoes created where surface water is heated to boiling temperatures by magma below. There are ten thousand such geothermal features there, more than in the rest of the world combined. Geologists estimate that magma lies just two miles below the surface. We have geology to thank for Old Faithful and the other volcanic features we enjoy at America's first national park. Unfortunately for sightseers, there are no scenic, volcano-shaped peaks remaining because they collapsed inward as magma beneath was erupted.

Because of shifting magma below the park, Yellowstone is one of the most seismically active places on planet Earth. Anywhere from one thousand to three thousand earthquakes rattle the park each year, although most are too small to be felt. The largest earthquake in the park's recorded history occurred during the night of August 17, 1959, at the west entrance to the park near West Yellowstone, Montana. Most people don't realize that a small part of the park lies in Montana and Idaho, even though Wyoming claims most of it.

Called the Hebgen Lake earthquake, this magnitude 7.5 quake—it was huge—caused a large landslide (80 billion tons of rock) to travel down the south side of the Madison River valley where it then traveled four hundred feet up the other side, creating hurricane force winds in the process. Unfortunately, the Rock Creek public campground was in the path of the landslide, and twenty-eight campers lost their lives. The slide buried a mile-long stretch of the river and adjacent highway, creating Quake Lake behind it. The highway had to be rebuilt on higher ground, and at one location as I drove along I could see the old highway and follow it until it disappeared beneath the water of the lake. Comparable in strength to the 1906 San Francisco earthquake, this example of Earth movement produced fault scarps (cliffs) up to twenty feet high that are still visible today. The Hebgen Lake earthquake was the strongest earthquake ever recorded in the northern Rocky Mountain region and the third largest in the continental

Typical scenery while doing exploration in Montana—
Granite Peak in the Tobacco Root Mountains, 1979.

United States. It was so strong it altered Yellowstone's plumbing system; within this slumbering supervolcano, new geysers were formed, and others dried up.

While I'm talking about Yellowstone, I'm reminded that I went to college with a person from nearby Bozeman, Montana, by the name of Cliff Montagne. While growing up, he spent a lot of time at Yellowstone, so his nickname at Dartmouth was YP Montagne (short for Yellowstone Park). Geology had even influenced what we called one of our friends!

While in the Absaroka Range, we reexamined the Emigrant Peak property that we had discovered the previous year. What a gorgeous day it was with up to ten inches of fresh snow at the eight-thousand-foot elevation that contrasted sharply with the clear, bright blue sky. There I was, admiring the scenery again! In spite of the snow, we saw lots of gossan (iron-stained rock from which minerals had been leached [removed]) and silicification (silica replacing minerals in rocks) and some molybdenite—all encouraging exploration signs. We decided we'd come back after the snow melted and do more work.

One of my visits to Emigrant Peak coincided with the nearby Livingston, Montana, rodeo, so of course my field assistants and I had to attend. I think it was the first one I ever attended and probably the first for my field assistants. We enjoyed the rodeo and fireworks celebration and got a good feel for the town. Livingston is a friendly community right on the Yellowstone River where the interstate highway gets a running start on the Continental Divide to the west.

Dividing my time between two states and among seven field assistants was a constant challenge but one I enjoyed. I wanted to be sure that each field team received an equal amount of my coaching, teaching, and training time. These field assistants were all eager to learn and hardworking, and I hope each one found his/her experience with me that summer to be positive. Even though some of them spent only three months with me, I still remember them fondly. I regret that I have lost track of many of them.

By midsummer, we were getting ready to drill at East Goat Mountain, so one day Janet accompanied me to Deer Lodge to look for road access to get us closer to the property before we had to start hiking. Then in late July, Randy Moore, Larry Campbell, and I hiked into our claim block and camped for the night; we were scoping out drilling locations. It was fun sitting around a campfire with the stars blazing above. I wanted to give Randy and Larry a break from Liver Peak and an opportunity to see another molybdenum prospect.

I then took rangers from the US Forest Service's Deerlodge National Forest to our proposed drill site to secure their final blessing on drilling, which they gave us. Finally, in early August we were ready to start core drilling. Janet accompanied me again to Deer Lodge, the staging area for slinging—by helicopter—the drill rig and field camp into the mountains. The helicopter lifting off with the drill rig hanging beneath from a steel cable was quite a sight, but everything went like clockwork. Janet even rode in the helicopter with us. By late September, we had completed drilling 1,338 feet of core in three exploratory holes, so we brought the helicopter back to demobilize the drill rig and the camp. At the same time, we selected a drill site for next year's drilling season; this year's drilling had been encouraging.

A curious thing happened at East Goat Mountain. Since Armco Steel was Noranda's joint-venture partner in the Idaho-Montana molybdenum exploration program, in late August I gave their geologists a tour of Liver Peak and East Goat Mountain. They were thrilled with our results, as I knew they would be. However, not until we had climbed an arduous trail up East Goat Mountain did I learn that one of the Armco geologists had a prosthetic leg! Amazingly, he handled the hike well and never complained.

Later that summer, Janet accompanied me on another trip, this time to south-central Idaho. On the way there, we stopped at Torrey's Cabins in Clayton and had dinner with our friends Phil and Val Johnson from Thompson Creek days. We were happy to see one another. Upon arriving at the Roaring River campsite of Dave Hyde and Rick Bellows the next day, Janet and I hiked with them up a strenuous five-mile route to a prospect we had identified from a literature search. Even though it was a difficult hike, we were eager to find out what was beyond the next bend. There's that geologist's curiosity again. Unfortunately, the prospect was not worth pursuing, but the scenery was beautiful.

That summer, there were too many other trips to south-central Idaho to remember; there was a lot of country there to be prospected. Consequently, I spent a lot of time on the road and in the air in 1979, traveling between Idaho and Montana. I often wondered how Janet put up with my being away so much. But she realized how much I loved my work, and she was willing to share me with it. It helped that I was able to take her with me on some adventures.

Long-distance driving was just part of the job. One day I drove about two hundred miles from Missoula to Lake Pend Oreille in northern Idaho where I examined a molybdenum prospect. I returned home the same day, arriving around eleven o'clock that night. The next day, I got up early and drove about 150 miles to Virginia City, Montana, where I looked at an old mine in the Tobacco Root Mountains.

It occurs to me now that I should have recorded and tallied up the mileage I drove each year between Idaho's northern panhandle to the Boise Basin in southern Idaho, to the eastern extent of the Belt Series rocks in central Montana. Although I drove a lot of miles, to this day I still enjoy driving; it gives me a joyful opportunity to contemplate the geology outside my window as I drive along. I resonate with Ian Frazier's comment from his book, *Travels in Siberia*: "In America we love roads. To be 'on the road' is to be happy and alive and free. Whatever lonesomeness the road implies is also a blankness that soon will be filled with possibilities. A road leading to the horizon almost always signifies a hopeful vista for Americans."[1]

All this driving provided occasional moments of unanticipated excitement. On one of my many drives across northern Idaho, I vividly recall completing a full, unintended, 360-degree turn in

---

1.   Ian Frazier, *Travels in Siberia* (New York: Farrar, Strauss, and Giroux, 2010).

my lime-green Ford Bronco while traveling down Interstate 90 at sixty miles an hour! This was a snowy, wintery day with black ice; fortunately, I did not hit anything or anybody—so I just kept on driving. It's actually too bad I didn't hit a guard rail so that I could have gotten a new vehicle. I never did like the outlandish color of that Bronco.

As a result of Dave and Rick's prospecting in northern Idaho, we eventually acquired a mining lease on seventy-three mining claims covering the Vulcan molybdenum property on the east shore of Lake Pend Oreille. I had taken a preliminary look at this property earlier in the year and thought it held some attraction. Dave and Rick's follow-up work supported that conclusion. The location was gorgeous, overlooking the lake; I especially enjoyed working there because the terrain reminded me so much of northern New Hampshire with its large expanses of forests, numerous lakes, and fast-flowing rivers. Dave, Rick, and I geologically mapped and sampled the Vulcan property's molybdenum- and tungsten-bearing stockwork-quartz veins that cut intrusive rocks and Belt Series metasedimentary rocks. We wanted to be ready to drill there the next year. After completing fieldwork, we put our heads together and developed a conceptual geologic model hypothesizing that Cominco American's previous shallow drilling there indicated the potential for ore-grade material lying at depth.

While exploring at the Vulcan property, we came across and acquired the nearby Chilco Mountain property, another stockwork molybdenum prospect in Belt Series rocks that was similar to Liver Peak. Even though three other companies had explored the property since its discovery in 1972, no intrusive rock had yet been found. Sound familiar? We acquired a lease on the property from Union Carbide Corporation and drilled it the following year.

All of this traveling took me to many small towns in Idaho and Montana like Lowman, White Sulfur Springs, Livingston, and Ennis, and I stayed in many now-forgotten motels. I ate my meals at local restaurants and cafes, most of those also lost to memory. I do, however, remember eating once at a restaurant in Helena, Montana, where I enjoyed beef tongue and spinach, a meal not commonly seen on a menu. My first taste of fried catfish was when I was on a field trip in Missouri, and it was delicious. Later, when I worked in the desert southwest, I developed a love of Mexican food. There was an especially good Mexican restaurant in Socorro, New Mexico, a town that I frequently passed through when doing exploration in the northern part of the state. I quickly discovered that the best Mexican restaurants were usually found on the other side of the tracks. When I worked at Thompson Creek and stayed at Torrey's Cabins, we geologists had the luxury of cooking for ourselves; fortunately for me, some of my field assistants were pretty good cooks. We often enjoyed great backyard barbecues. But good food was not the primary focus here; molybdenum was!

11. DISRUPTIVE ERUPTION

Even though it began slowly, the year 1980 promised to be busy and exciting—during the summer, we would drill four properties in two states: Liver Peak, East Goat Mountain, Chilco Mountain, and Vulcan. I didn't mind the slow start though, because I put the time to good use in the office, massaging data and making detailed plans for drilling. Lots of effort was required to solicit, review, and award drilling contracts, especially for four distinctly different properties. We exploration geologists loved the drilling phase of property evaluation because that's where the most excitement was—it gave us a chance to exercise our four-dimensional thinking skills.

During the winter, I did get out of town once when I flew to Denver to help finalize the lease agreement with Union Carbide Corporation on their Chilco Mountain prospect. Thank heavens Noranda had a landman on staff in Denver who did most of the property acquisition work for us. That wasn't my cup of tea.

In February, I went underground again, this time at the Galena and Sunshine mines in Idaho's northern panhandle. I had great tours of these operations and learned about the sad history of the Sunshine Mine. Though one of the largest silver producers in the United States, this mine had experienced a blemished past when on the morning of May 2, 1972, a fire started in the underground workings of the mine. Before it was over, ninety-one miners died of smoke inhalation or carbon monoxide poisoning, making it one of the worst mine disasters in the United States. As a result of this fire, all miners in the United States are now required to wear self-rescuers to prevent carbon monoxide inhalation.

That spring I once again exercised my civic right—and duty— to express my opinion by preparing testimony in opposition to the federal government's attempt to prematurely, in my opinion, create a Big Hole Wilderness Area in southwest Montana. Janet, Tom Butler (another Noranda geologist), and I drove to Dillon, Montana, where all three of us presented written testimony to the House Subcommittee on Public Lands, which was holding public hearings on the subject. The thrust of my argument was that the oil and gas potential of the area had not yet been evaluated, so it should not be "locked up" in wilderness designation before that evaluation was done. Some of my non-mining Dartmouth friends who were on the other side of that issue were also there. Interestingly enough, on our return home, we saw a large oil-drilling rig just five or ten miles from the proposed wilderness boundary near Wisdom, Montana. The subcommittee ended up not declaring it a wilderness area.

In late April, Geoff Snow convened another session of Noranda's Witchers and Dowsers Society, this time in Cornwall, England. The purpose of the trip was to examine this famous tin-mining locale and an ophiolite (a sequence of rocks that was once part of the ocean floor) along the English Channel. After flying into Heathrow Airport and boarding our tour bus, we stopped at Stonehenge, the famous Bronze Age earthworks of large stones standing upright in a circle. In those days, there was no fence around this archeological site, so I walked right up and inspected the rocks. In case you always wanted to know, the rocks are diabase, rhyolite, volcanic ash, micaceous sandstone, and greenish sandstone. Stonehenge is definitely an engineering tribute to those who built it some 3,100 years ago.

Although I didn't realize it at the time, the town of Bath is only about twenty miles west of Stonehenge. As you may recall, the New Hampshire town in which I grew up is called Bath and was named after England's 1st Earl of Bath. I wish now I had visited the original Bath.

From Stonehenge, we traveled south to pick up Nick Badham, a renowned economic geology professor at the University of Southampton, who served as our local guide. We then drove to Cornwall—the most southwestern part of England—through what must be typical English countryside: lush, green fields with a myriad of hedgerows, flowers in bloom everywhere, and lots of cattle and sheep, including newborn lambs. We stayed in roadhouses or small hotels, and I found the people to be extremely friendly and welcoming. They are also neat as a pin—everything was clean and tidy. I especially enjoyed the welcoming and family-oriented pubs where I was introduced to the Ploughman's lunch, the gastronomic icon of England consisting of a cold meal of cheese, pickles, bread, and butter, usually served with a pint of beer. And I loved the fish and chips, too.

To get to Cornwall, we passed through rural County Devon, famous for its extensive moorlands and generally hilly terrain. As we drove along, I wondered if we would be seeing Sherlock Holmes and any of the hounds made famous in *The Hound of the Baskervilles*.

The Land's End Peninsula of Cornwall has a long-standing mining tradition, and its coastal cliffs provide good rock exposures for geologists. Prior to 1700, tin was the only metal of interest here. Tin was first mined from streambeds (like placer gold is), probably as far back as the Bronze Age (2150 BC). Then, beginning in the Middle Ages, tin was mined from quartz veins that cut the local granite. During the Industrial Revolution, copper took on greater importance, so much so that by 1800, Cornish copper deposits were the primary source of copper for Europe and throughout the rest of the civilized world. However, by the end of the 1800s, other sources of copper had been discovered, such as in the Keweenaw Peninsula of Upper Michigan, so tin mining again became the primary metal being sought in Cornwall. Cornish tin was renowned for its purity, and mining continued there well into the mid-twentieth century.

Remnants of the heyday of Cornish mining were evident by the presence of crumbling, brick smokestacks and huge, stone engine-houses that to me resembled Irish castles. Sitting high on the cliffs overlooking the English Channel, these stately structures provided excellent photo opportunities. What breathtaking scenery welcomed us here where the English Channel meets the wide-open Atlantic Ocean.

Most mining was done by underground methods, and many of the underground workings went to depths of two thousand feet. Consequently, they were below sea level, and huge pumps were needed to keep the workings dry. The engine-houses that I saw dotting the landscape housed gargantuan steam engines that powered the water pumps. Some of the underground workings went as far as three-quarters of a mile out beneath the sea. On occasion, in their eagerness to get every bit of ore, the miners would inadvertently mine up into the sea bottom, thus flooding themselves out. The Cornishmen became experts at removing water from mines, and their expertise was in demand worldwide.

In our traverses along the cliffs, we often saw mine workings where they came to the surface; and on the rocky beach, I collected cobbles of quartz veins containing tin-tungsten minerals. In places where the granite had been highly weathered, kaolin clay had been mined from surface pits; this was the source of the famous English china clay. At low tide, we walked along a causeway a few hundred feet out into the English Channel to St. Michael's Mount to see the tin-bearing granites there. At high tide, the outcrop, including a castle at the top, becomes an island.

Toward the end of the trip, I was thrilled to see my first example of a remnant of oceanic crust at Lizard Head, also right on the coast. Called an ophiolite sequence, this is a classic example of oceanic crust that had been thrust upon continental crust by plate tectonic movement and thus represents a tectonic plate boundary. How exciting!

I loved the names of many of the places we passed through in Cornwall: Bodmin Moor, Dartmoor, Land's End, Cligga Head, Goonbarrow, Goonhilly Downs, and Godolphin. We passed by the town of Poldark, also the name of a popular British television drama series in the mid-1970s that was set in Cornwall. As we stopped in the town of Penzance, we were reminded of the musical *Pirates of Penzance* that was popular at the time. Bob Hodder, who accompanied us on this trip, then quipped, "If we find any pyrite here (an iron sulfide mineral), they can make a musical titled, *Pyrites of Penzance!*"

On this trip, we came within ten miles or so of the town of Dartmouth; my alma mater, Dartmouth College, was named for the 2nd Earl of Dartmouth. We didn't get to go to that town either! Darn! Before flying home, we spent the last two nights in Oxford, which provided me with an opportunity to tour London and see its famous sights. The trip to Cornwall was an excellent opportunity to see some unusual geology and gain some additional cultural insight.

When back in this country, we resumed drilling at Liver Peak as soon as the weather permitted. Randy Moore once again served as the onsite geologist, responsible for core splitting and geologic logging, mapping, and supervising the drillers. We drilled our first drill hole of the season, designated LP-5, to about four thousand feet and intersected good-looking mineralized rock in the Upper "Ore" Zone. Deeper in the hole, we intersected igneous rock—also heartening news.

Then, partway down an existing drill hole, we tried an innovative drilling approach by installing a down-hole motor on the drill bit—a Dyna-Drill, it was called—so we could drill at an angle off from the original vertical borehole. This drilling method had never been tried before in a "hard rock" setting; this was technology transfer from oil field exploration. However, the rock was too hard for the drill bit, and the method was not successful. We twisted the drill bit off in the hole more than once and got the drill stem (drill rods) stuck in the hole as well. I spent a lot of time on the telephone with the drillers and driving back and forth to Liver Peak from Missoula trying to find a way to make this drilling method work, but to no avail.

What a frustrating year of drilling it ended up to be, not nearly as good as 1979. In addition, our calculations from the latest drill-core assay results reduced the possible tonnage of ore from four hundred million tons to one hundred million tons. However, on one of my trips to Liver Peak that summer, I did see four bears and two deer. At least that was exciting.

On the afternoon of May 18, Janet, Deb, and I were passing through the Bitterroot Valley on a Sunday drive when we saw unusual yellowish-brown clouds coming over the mountains from the west. We wondered aloud what that was all about. By the time we arrived home, gray ash particles were falling from the sky. It turns out this was the leading edge of the volcanic ash cloud coming from the cataclysmic eruption of Mount St. Helens in southwestern Washington. When all was said and done, we had about half an inch of gray ash covering everything at our house, so I collected jars full of samples as souvenirs. I still have some today. The ash was so thick in the air that the city streetlights came on during the day, and it was suggested that everyone hose off the roofs of their houses, which I did. My daughter reminds me that she got days off from school as a result of this geologic event. We were much more fortunate than those fifty-seven individuals living in southwestern Washington who experienced the full disruptive force of the eruption and lost their lives.

That drive through the Bitterroot Valley affords a beautiful view of the glaciated terrain of the Bitterroot Range. The scouring action of glaciers on up-faulted mountains is responsible for much of our spectacular western mountain scenery. The Lewis and Clark expedition must have enjoyed this same scenery when they explored this part of Montana; we have a reminder of them at Lost Trail Pass where US Highway 93 crosses from Montana into Idaho atop the Bitterroot Range. The pass derives its name from the fact that Lewis and Clark became lost there on September 3, 1805, and had trouble finding a way through the mountains.

In June, two of my new summer field assistants, Bill Beyer and Rick Bellows, and I went to Chico Hot Springs at the base of the Absaroka Range to revisit the Emigrant Peak prospect we had identified last year and to look at other areas as well. We set up a camp trailer at Chico Hot Springs as our base of operations and enjoyed experiencing a violent thunderstorm the first night there. What raw power came from that pyrotechnic spectacle: hydraulics! voltage! percussion! The next day, as I was hiking across a steep, hard-packed snowfield on the slopes of Emigrant Peak—and without crampons or ice axe—I slipped and slid out of control for a few tens of feet before catching myself on a small evergreen tree. It's a good thing I stopped, because it was a long drop to the bottom.

The author holding a length of drill core from
East Goat Mountain, Montana, 1980.

The only harm done—aside from scaring myself half to death and receiving a few scrapes and scratches—was the loss of my wristwatch when it became caught on a snag. It's still up there somewhere.

Now we had started drilling again at East Goat Mountain, so one day I went there to check on drilling progress. This time I flew in a Jet Ranger helicopter that I had contracted from Missoula. I flew there in the morning, examined the drill core, and flew back home in the afternoon. What a spectacular, if not unusual, commute. We were getting good-looking core out of the drill holes at East Goat Mountain, and that continued until the end of drilling in September when the weather turned bad and we pulled out.

At the start of the summer, I had assigned Pat Wotruba, another new field geologist this season, to the Chilco Mountain and Vulcan projects in northern Idaho where drilling would soon begin. He and I scouted out drill sites and found a local cat-skinner to build drill roads and drill sites for us. By the end of June, we were ready to start drilling at both properties. When I visited Pat one day later in the summer to check on drilling progress and status of his detailed mapping at Chilco, I was pleased to see he had discovered nice outcrops of molybdenite in Revett Quartzite of the Belt Series rocks. That was encouraging.

After examining drill core from Chilco Mountain and Vulcan with Pat, I then drove by myself south to Lowman, Idaho, arriving at eleven at night. The next morning, I met Roger and Mary Jo Kuhns for breakfast. They were a young, newly married couple whom I had hired as field assistants to explore the Idaho Porphyry Belt of the Boise Basin. Roger Kuhns is the person who introduced me to computers. He showed up for work at the start of the summer with a new-fangled Apple II computer. Not until two years later did we have one in the office; and we used it only for report writing, and sparingly at that. The technology wasn't available at that time to use a computer for geologic research purposes. The three of us spent the day taking a six-mile hike to look at a prospect. I don't remember anything about the property except that it had no merit; most property evaluations have this result, so this was not unusual.

Roger and Mary Jo worked out of a travel trailer set up near Lowman, and I visited them often that summer to review their reconnaissance results. We once took an eight-mile hike—one way—up Red Mountain near Lowman, where we discovered potassic alteration of the rocks along with some gorgeous scenery. Once again, the prospect had no merit. From Lowman, I drove to Torrey's Cabins for dinner with Phil and Val Johnson and spent the night in Salmon, Idaho. I was home in Missoula the next day.

In mid-July, I flew to Billings, Montana, and then drove south to the town of Big Timber to examine a property in the Beartooth Mountains. This was my first trip to that mountain range. Bill Beyer met me there, and we hiked eight miles along a logging road to look at the Gold Hill property, but it wasn't too exciting.

What *was* exciting, however, was to be close to Stillwater Mining Company's two mines where platinum and palladium were being extracted from the 2.8-billion-year-old Stillwater Complex. This geologic feature is a layered succession of mafic and ultramafic igneous rocks rich in iron and other heavy metals. Significantly, these mines are the only platinum-palladium mines in the United States, and they contain the highest grade platinum ore being mined anywhere in the world. Alas, platinum and palladium were not what we were in search of. Perhaps they should have been.

Later in the summer, I invited my summer field assistants who were in town to join Janet and me at our house for a roast turkey feed one Sunday night: Bill Beyer, Jeff Tepper, Rick Bellows, and Roger and Mary Jo Kuhns attended. Eating a home-cooked meal was a real treat for these people who spent so much of their time eating in restaurants. And it gave Janet and me an opportunity to get to know them better on a personal basis. That summer was hectic for me, keeping track of each field crew, monitoring and evaluating their work, and scheduling their time, so it was nice to be able to sit and relax with them. I found that I enjoyed managing the crews and did not have a problem giving up doing all the fieldwork myself; that is, I was willing and able to delegate. Not all exploration geologists can do that.

Until late October, I was on the road and in the air much of the time, traipsing all over Idaho and Montana, going between our drilling projects and raw prospects that we were evaluating. Frontier and Western Airlines got to know me pretty well. Drilling at Vulcan ended in late September, but Chilco Mountain and Liver Peak drilling continued into December. At Chilco, our last drill hole was a deep one; it bottomed at 3,500 feet. On one trip there, my field assistants and I stayed up until midnight discussing what they had found in that deep drill hole and what it might mean for the future of the prospect.

From November 9 to 11, Noranda held a "Fall Rendezvous" for its geologists at the company headquarters in Denver. This was not a session of the Witchers and Dowsers Society since it did not include a field trip. Instead, the meeting provided an opportunity for all Noranda geologists to get together and share geologic lessons learned, exchange expertise, and develop teamwork. At the meeting, we learned about skarns (limestones that have been altered to silicate minerals by hydrothermal solutions), precious-metal deposits, and exploration management, and then we all took a technical-writing course. That course was badly needed since most geologists are not naturally good writers. Noranda geologists attended from all the district offices: Anchorage; Missoula; Reno; Tucson; Denver; Rhinelander, Wisconsin; and Lebanon, Tennessee. Each office was comprised of four or five geologists, and as you can imagine, this meeting was a large and boisterous gathering.

Shortly after returning home from the Fall Rendezvous, I was greeted with the joyous event of the birth of our daughter Meaghan on November 17. What a cutie she was! *How is her arrival going to affect my traveling ways?* I wondered. I already knew how she would affect my sleep; her eruptions of crying kept both Janet and me up for many nights. Her birth did not slow us down, however, at least not in the near term. In early December, Janet and I attended the Northwest Mining Association Annual Convention in Spokane. If I recall correctly, Roger and Mary Jo Kuhns (the newlyweds) took care of Meaghan in our absence. Deb was old enough to stay at a friend's house.

12. TO THE DESERT HERE WE COME

In January 1981, Geoff Snow asked me to take the job of district manager of Noranda's southwest district office, located in Tucson. Thinking that a good New England boy would not like the hot Arizona desert, I respectfully declined. Shortly thereafter, he asked me again, and being fairly astute, I recognized he really wanted me to do it, so I said yes. It was obvious that I had a good case of wanderlust anyway, so why not take this on as a new adventure. Besides, by this offer, Geoff had expressed his confidence in me both as a geologist and as a leader, so why not take on the increased responsibilities.

So in late February, Janet, Deb, and I flew to Tucson to see what we were getting ourselves into. As you'll see later, it turned out to be a good tour of duty. We must have left Meaghan—two and one-half months old—in the care of someone in Missoula because we did not take her with us. We started making plans to move in early summer after Deb finished her freshman year of high school in Missoula. I would start my new job sooner than that, however.

Back in Montana, our drilling at Liver Peak continued through the winter. Randy Moore was still running the project and doing a good job. That spring, I made occasional visits to Thompson Falls-Liver Peak to see how things were going. On one of my last trips there, I saw about one hundred bighorn sheep grazing in a meadow. Ho-hum, just another day at work for an exploration geologist!

By the end of this year's Liver Peak drilling program in October, Randy and his drill crew had drilled four additional core holes into the Upper "Ore" Zone and better defined the size and grade of this zone. Even though the grade of ore was good (averaging 0.26 percent MoS_2), the identified tonnage was now reduced to only thirty million tons rather than one hundred million. Randy also drilled a fifth hole into the Lower Zone but encountered only low-grade mineralization.

Using what I had learned from the Colorado School of Mines' economic evaluations course taken a few years earlier, I did some financial analysis of our assay results. Unfortunately, my calculations revealed that the small size of the Upper "Ore" Zone and the low grade and extreme depth of the Lower "Ore" Zone made mining economically infeasible. After drilling a total of 30,019 feet of core in ten holes and spending four years in exploration (in which I was involved), Noranda dropped the property. The low market price of molybdenum, low grades of molybdenite, and the small tonnage available convinced us that the property did not fit the working definition of ore. So much for all that optimism we experienced in 1979!

Did you notice how the anticipated tonnage of ore decreased continually as we gathered more data and learned more about the Liver Peak property? Starting with a possibility of four hundred million tons in 1979, the tonnage decreased to one hundred million tons at the end of 1980 and ultimately to only thirty million tons in 1981. This phenomenon, however, is not atypical of exploration projects; the success rate in minerals exploration is very low. Of the few prospects that reach the drilling stage, only a small percentage ever become operating mines. We geologists must be perpetual optimists about what we may find; otherwise, we would not be in this business.

Working in the great Northwest had been a true adventure, but the time had come to move on to my next adventure. April 6, 1981, was the first day of my new job as manager of Noranda's southwest district. The day was filled with excitement and not a small amount of anxiety. The excitement came from the promise of new geology to see and sights to experience. The anxiety came from the realization that I was now in charge. I was solely responsible for the successes and failures of Noranda's exploration efforts in the great Southwest, which encompassed Southern California, Arizona, New Mexico, west Texas, and eventually, northern Mexico. Working out of that office were seven full-time geologists, numerous temporary field geologists, one technician, one draftsman, and a secretary. The experienced, full-time geologists didn't require much of my supervisory time, thank heavens, but ensuring the smooth operation of the office did—planning, budgeting, scheduling, personnel management, hiring and firing—all those things required to maintain a highly functioning organization. At only thirty-two years old, I was the youngest of Noranda's district managers and determined to be successful.

Not surprisingly, the weather in Tucson that first day was gorgeous; temperatures were in the eighties with perfectly clear skies. *Hmmm, maybe this position won't be so bad after all.* One of my immediate tasks was to bring myself up to speed on the exploration projects already underway. So during my first week there, Bob Page, a fellow Dartmouth graduate and Burlington, Vermont, native, took me to look at three copper prospects he had discovered on the Papago Indian Reservation (now called the Tohono O'odham Indian Reservation) west of Tucson. He was exploring for porphyry copper deposits like those I had visited on field trips in Arizona with the University of Western Ontario geology department. Thanks to those trips, I already knew what the favorable geology would look like.

It was with Bob that I learned an important lesson for a geologist—while working in the southwest, don't pick up the head of your rock hammer with bare hands. In that hot Arizona sun, a rock hammer sitting for not very long will become dangerously hot to the touch. I am convinced that a different sun watches over Arizona, a much hotter one than anywhere else. Even though it looks like any sun anywhere, the difference is the air temperature becomes immediately hot as soon as the sun appears above the horizon each morning and only gets hotter as the day progresses.

Because Noranda still had a molybdenum exploration joint venture in the Northwest with Armco Steel, we reported to them regularly on progress, and one of those progress meetings was scheduled for the following week. That didn't give me much time in the field with Bob because I had to fly back to Missoula to prepare for the meeting and then fly to Armco's headquarters in Middletown, Ohio, for the meeting. Following a good one-day session with Armco's geologists at which I presented our drilling results from the four properties in Idaho and Montana and exploration plans for the summer, I returned to Missoula to be with my family for a few days. That was refreshing.

The following week, I flew to Noranda's headquarters office in Denver to participate in my first Noranda district managers' meeting. I would attend many of these two-day meetings during the next four years at which we seven district managers developed exploration strategies, established budgets, determined staffing requirements, and celebrated our exploration successes. The most impressive exploration highlight that excited us all during this meeting was the drilling of ore on Admiralty Island, not far from Juneau, Alaska, where geologists from Noranda's Anchorage office had discovered mineralized outcrops in 1975. Called Greens Creek, when exploration was complete and mining began in 1988, this substantial silver-gold-zinc-lead mine became the fifth largest silver producer in the world.

Once back in Tucson, I continued learning about exploration projects already underway in the district: War Eagle/Gladiator near Jerome, Arizona (copper); Lakeshore, Arizona (copper); Greater Grants Joint Venture in New Mexico (uranium); molybdenum reconnaissance in New Mexico; and Four Metals, Arizona (copper). All this new geology made my head spin.

Earlier that spring, Janet and I had purchased a house in northwest Tucson on Sahara Palms Drive. Since Janet was still in Missoula, my assignment during nonworking hours was to supervise the construction of a swimming pool and surrounding block wall on our property. Both turned out beautifully, and we enjoyed about five years of backyard fun as a result. We moved into our new home on June 8, 1981, which, at 108 degrees, turned out to be the hottest day of the year. Nevertheless, we adjusted to the heat and got along nicely in Tucson. Coming home from work at night that first summer, I immediately took a dip in the pool. I even let the rest of the family enjoy it. Meaghan learned to swim there, and Deb and her friends could often be found lounging around the pool. I felt like a king—I had my own swimming pool! We had great neighbors, too, some of whom remain good friends. A javelina even paid us a visit one day, running through our side yard, and at night we could hear coyotes catching rabbits; their screams of death were haunting, however. Of course, Arizona's low humidity was wonderful, too. One day in Tucson, it measured only 2 percent. That made it nice for working outside—sweat never falls from your forehead onto your glasses because it evaporates first.

In July, I flew to El Paso, Texas, and then drove to Ruidoso, New Mexico, where Dave Jones and Henry Truebe, geologists from the Tucson office, were conducting molybdenum reconnaissance in the Sierra Blanca Mountains. Dave was another Dartmouth graduate whose career blossomed much later when he discovered Mexico's largest gold mine—Los Filos—that was put into production in 2010. Henry Truebe was a Colorado School of Mines graduate who became well respected for his knowledge and expertise in caving and mineral collecting.

I was familiar with the Sierra Blanca part of the world since I had worked there back in 1975, and it was good to be back. One town there, Ruidoso, is a mountain-resort community nestled in thick, cool woodlands that provide respite from the desert heat. Another town, Lincoln, is a tiny place that sits in a valley and has nowhere else to go.

Lincoln is one of the more notable places in the Sierra Blancas because of two things. First, it is in the heart of the Lincoln National Forest, the home of Smokey the Bear—or Smokey Bear, as the US Forest Service prefers. As the true story goes, in 1950, a badly burned bear cub was discovered clinging to a tree during the raging Capitan Gap forest fire. He was rescued and cared for and went on to achieve lasting fame as the representative for the Forest Service's campaign, "Only you can prevent forest fires." Today, Smokey Bear is one of the most widely recognized symbols in the world.

The second reason for Lincoln's fame is this is where the Wild West outlaw Billy the Kid rose from obscurity to become a national folk hero. Known far and wide in New Mexico Territory for stealing horses and cattle, killing people, and generally thumbing his nose at the law, this twenty-one-year-old gunfighter was finally caught in April 1881 and was being held in Sheriff Pat Garrett's jail awaiting hanging for killing the previous sheriff. Ever resourceful, Billy overcame and killed his two jailers and jumped from the second-story balcony of the Lincoln County courthouse to his horse below. Lore has it that while riding down Main Street, he yelled, "Three cheers for Billy the Kid!" Three months later, this iconic western bad guy was shot and killed by Sheriff Garrett. The spectacular jailbreak is celebrated annually during Lincoln's "Billy the Kid Pageant" where the historic event is recreated. I never did get to see it, however.

While in the area, I had dinner with my old friend from Truth or Consequences, Ralph Forsythe, who lived nearby; we dined at the Inn of the Mountain Gods on the Mescalero Apache Indian Reservation. In addition to enjoying these personal diversions, Dave, Henry, and I examined some good-looking geology in the Sierra Blanca Mountains. We didn't find anything worth staking on that visit, but our reconnaissance would continue.

As part of my professional development, I flew to Ann Arbor, Michigan, for a two-day course at the University of Michigan titled "Management for New Managers." I was now in a senior management position, so this was educational time well spent. They don't teach management courses in geology school! Later on, I often referred to my class handouts and notes for guidance as I dealt with knotty managerial problems.

At the end of July, I accompanied Mike Donnelly, another geologist from the Tucson office, to Wickenburg and Jerome in central Arizona to look at massive sulfide properties: the Vulture mine, Swastika mine, Crown King, and War Eagle/Gladiator. These inactive mines had been developed on massive sulfide deposits like the one I had examined on a Bob Hodder field trip on the Maine coast eight years earlier, so I felt comfortable with their geology.

These Arizona properties were in "elephant country" because they were in the same geologic environment as the old United Verde mine, a huge deposit of volcanogenic massive sulfide ore

(Precambrian, 1.7 billion years old) containing high grades of copper, silver, gold, lead, and zinc. In the Precambrian eon, this part of Arizona had been underwater, and the ore at United Verde was produced by volcanic eruptions onto the seafloor. The United Verde was the geologic model for the kind of copper deposit we hoped to find in this part of Arizona.

Initially, United Verde was an underground mine with eighty-one miles of mine workings; it was later exploited by open-pit mining methods. Mining operations were conducted there from 1883 to the 1960s. Interestingly, the ore was so rich in sulfide minerals that it was subject to spontaneous combustion when exposed to air; underground fires burned unabated for many years. While mining from the open pit, huge, coal-fired steam shovels would often dig into burning ore and load smoking rock into the rail cars. One such shovel had originally been built for digging the Panama Canal.

The United Verde mine and adjacent United Verde Extension mine were great, great mines in their day, and others like them were a worthy exploration target. The profits from these two mines were solely responsible for putting the Phelps Dodge Corporation—a major US mining company at the time—in business.

At its peak in 1929, the nearby mining town of Jerome hosted fifteen thousand people and was the largest city in Arizona. Jerome sits perched and precariously anchored on a thirty-degree slope on Mingus Mountain two thousand feet above the Verde River Valley. Today, Jerome is barely a remnant of its former self, consisting of only a few hundred residents, many of them artists. These words about Jerome, written by folksinger Kate Wolf, are appropriate: "She once was a miner's city, now the ghost of a dying town, But there's a fire burning bright in Old Jerome."[1]

On this trip with Mike, I learned another valuable lesson for geologists working in the desert—always carry a pocket comb. Combs are essential for removing cholla cactus from your clothes or body. The jumping cholla, as they are commonly known,

---

1. Kate Wolf, *Old Jerome* (Forest Knolls, California: Another Sundown Publishing Company, 1983).

seem to "jump" onto you as you walk by them and are almost impossible to remove with your bare hands; once one cactus spine is removed, another one sticks to you. So a pocket comb can be used to easily flick off the offending spine without having to touch it. The spines sometimes attach themselves so strongly to the leather of your field boots that you have to use a pair of pliers to remove them; I know this from personal experience.

In early August, I flew to British Columbia for a weeklong educational tour of mines in the porphyry copper belt of that Canadian province; this was my fifth trip to British Columbia. Along with fellow Noranda geologists Bob Page, Eliseo Gonzalez-Urien, and Leslie Landefeld, I flew into Vancouver and then to Smithers, British Columbia, in the northern part of that beautiful mountain province. We were given excellent tours of huge open-pit copper/molybdenum mines: Endako, Gibraltar, Afton, Highmont, Ingerbelle, and Lornex. We also went to Noranda's Bell copper mine, which is scenically located adjacent

Mine shovel and haulage truck at the Ingerbelle
Copper Mine in British Columbia, 1981.

to Babine Lake. The late-summer weather and mountain scenery were gorgeous, and the geology was exciting. During the trip, I became acquainted with the British Columbia towns of Kamloops, Quesnel, 100 Mile House, and Prince George.

While returning home through Vancouver, my old friends from the University of Western Ontario, Harlan Meade and his wife, Linda, gave me a wonderful guided tour of their beautiful city; high mountains provide a beautiful backdrop to this city sitting right on the waterfront.

At the end of the month, I returned to New Mexico to monitor progress of Noranda's molybdenum reconnaissance program being conducted there by Dave Jones and Henry Truebe. With the two of them, I looked at properties near Bonito Lake and in the White Oaks District, again near Ruidoso in the Sierra Blanca Mountains. This lone mountain range in the central part of the state is the closest high-mountain country to Texas, and as we drove through Ruidoso, it seemed as though we saw more license plates from Texas than from New Mexico. From there we went a little north to the Gallinas Mountains where we were excited to find free-gold in outcrop, even though that was not what we were looking for.

A few days later, I went to Albuquerque for a meeting of the Greater Grants Joint Venture, of which Noranda was a partner. This corporate partnership was exploring for uranium on the Colorado Plateau. Noranda was providing funding for the project but no geologists. So far, the venture had not met with success, and shortly thereafter, we dropped out of the partnership. But since I was already in northern New Mexico, following the Greater Grants meeting I drove north even farther to the Taos area to look at molybdenum prospects around Eagles Nest, Red River, and Questa. This was familiar territory from my exploration days there in 1975, and again it was nice to be back.

One day after returning to Tucson, Mike Donnelly and I chartered a private plane and headed to Silver City in southwest New Mexico to look at a molybdenum property. It was a gorgeous day and excellent flying weather; however, as is most often the case, we found the property not worth pursuing. Later in the month, Mike and I again chartered a plane, this time to go to Albuquerque to meet with the owner of another molybdenum property, the Black Veil prospect in northern New Mexico. This one held more attraction.

As a result of these trips, I came to appreciate Albuquerque's geologic setting. This largest city in New Mexico is about 130 miles north of my old stomping grounds of Truth or Consequences, both right on the Rio Grande. Albuquerque lies in a dramatic setting, with the Sandia Mountains forming its backdrop on the east and the broad Rio Grande Rift stretching off for twenty-eight miles to the west. The steep, rugged, western face of the Sandias is an eroded fault scarp rising about five thousand feet above downtown Albuquerque. Faulting has been so dramatic here that the same Pennsylvanian (299–318 million years old) sedimentary rocks exposed on the west face of the Sandias can be found nearly twenty thousand feet down in the rift beneath the city. I wondered if the residents of Albuquerque knew they were living next to a major fault. This wasn't California, after all.

In October, I made another one-day trip to Albuquerque to negotiate a deal with the owner of the Black Veil prospect that we had taken an interest in a couple of months earlier. From there, I flew to Denver for a Noranda district managers' meeting where we worked on developing the budget for 1982, one of the less fun parts of the job. Back in Tucson afterwards, I spent a few days putting the final touches on my district budget before flying to Albuquerque again—I really became acquainted with that airport, having made six trips there by year's end—and then driving north to Taos. East of town, Dave Jones and I hiked through five inches of new, white, fluffy snow to look at the Wheeler Peak molybdenum prospect Dave had discovered in the glacially sculpted Sangre de Cristo Mountains, the southernmost range of the Rocky Mountains. Little did I know that during the next three years we would find more molybdenum prospects in the Sangre de Cristos that afforded us many return trips.

We were excited to find substantial molybdenite exposures at Wheeler Peak near the Wheeler Peak Wilderness Area;

consequently, we staked 125 mining claims. These claims covered an impressive stockwork of quartz-pyrite-molybdenite veins. At 13,161 feet in elevation, Wheeler Peak off in the near distance is the highest point in New Mexico. The day was beautiful as we hiked around, and I especially enjoyed the wonderful view about six thousand feet below us of the town of Taos, only about fifteen miles away. There I was, looking at the scenery again! The Native Americans certainly chose a scenic place to build their Pueblo de Taos. Almost unbelievable is this collection of reddish-brown adobe buildings, some five stories high, that was built about one thousand years ago; this is the oldest continuously inhabited community in the United States.

While hiking around in the Sangre de Cristos with our noses to the outcrops, we were giving every rock a chance to tell its story. Each rock does have a story to tell, granted, some more compelling than others. But no matter how insignificant a rock or outcrop may seem, the story it holds is worth knowing.

In late November, I flew to Spokane, Washington, where I presented a technical paper about the Liver Peak Project at the annual convention of the Northwest Mining Association. Even though Noranda was no longer exploring at Liver Peak, I was proud of the work we had done there and wanted to share with my peers what we had learned. This overrode the competition-driven desire for secrecy that sometimes accompanies exploration projects. I managed to work through my stage fright as I described the project to hundreds of my fellow professionals, and, all in all, I was pleased with the presentation.

After only nine months on the job as district manager, I was seeing success. Our reconnaissance efforts were turning up raw prospects, attractive properties were being identified, mining claims were being staked, and the district staff members were working together as a well-oiled unit. The big question was, would we be drilling next year?

13. HAZARDOUS DUTY

Since Mike Donnelly's Precambrian massive sulfide exploration program was showing some success, in January 1982, I accompanied him to central Arizona to see the key properties he had identified: Kay prospect, De Soto mine, and Blue Bell mine. The weather was gorgeous in the cool, high country around Jerome and Cottonwood as we took samples of copper, lead, and zinc minerals exposed there. These Precambrian deposits are also rich in pyrite, an iron-sulfide mineral that generally has no economic value. However, at one of these old mines, a local entrepreneur was crushing the pyrite from a waste pile into granular form and then bagging it for sale as a source of iron to make lawns green. You may recognize it as a generic form of Ironite. This enterprising soul was making money from what otherwise would have been waste rock. Mike was doing a good job identifying promising-looking prospects, so I left him to find one that we could drill.

My next event was far less enjoyable. I attended a district managers' meeting in Denver that turned out to be an unsettling affair because we managers were told to reduce our district budgets for next year by 25 percent. What a distasteful and difficult task. *Would we have to fire anyone?* we wondered. It turns out we did not; we were able to reduce costs by other means. Funding throughout the minerals industry was getting tight, and Geoff required that we address it head-on. We district managers did not foresee how "head-on" Geoff intended to be about this. This budget-cutting turned out to be a forerunner of more difficult times to come. What's that black cloud looming over our heads?

Two weeks later, I flew again to Denver to talk with Conoco geologists about their Jones Hill massive sulfide property in northern New Mexico. While prospecting one day, Dave Jones had stumbled across this property, and we were interested in conducting exploration there if we could negotiate a favorable agreement with Conoco. But try as we might, our negotiations were unsuccessful.

In February, I flew again to Albuquerque where Dave and I met with New Mexico Congressman Manuel Lujan. We continued to be interested in the Wheeler Peak molybdenum property in northern New Mexico, and we needed the congressman's help—the property was close to the Wheeler Peak Wilderness Area, and exploration was not allowed in wilderness areas. Even though our claims were not *in* the wilderness area itself, they were close enough that mining would be difficult. I argued that the wilderness area designation had been made without full consideration of the mineral resource potential of the area. My argument was duly considered and then forgotten by the powers that be. No changes to the wilderness designation were made.

As an exploration geologist, I interacted with all sorts of interesting individuals, from congressmen to environmentalists to lone prospectors. Some of these individuals were real characters, and my interactions with them were sometimes fascinating. What follows is a letter I received in April 1982, from an old prospector whom I had politely turned away after he had asked me to look at his property with him:

> *Dear Mr. Lamarre:*
>
> *I was surprised to receive your letter of Feb. 23 explaining your current position and limitations. Thank you for any consideration per my request for your special exploration assistance. I have written to over a dozen large mining concerns, and Noranda proves to be the most business-like. Perhaps you may take the time to review my position: the gold placer and lode operation I referred to is a site which has*

taken me seven years to pinpoint. A very old mountain man told me of a certain lost mine, explaining that we would die should we try to stake [claims on] the mineral there. For some reason, the old chap decided it was time to prove to all that he knew where a fabulous gold deposit lay—we were to give the mine to the government. Unfortunately, the man died before we could get to the site. After some time, and study, I realized that he had told me exactly where to find the mine. Incidentally, the mine is quite historic, being the first gold discovery by white men in the area.

I tried twice to get to the spot—it is very isolated, surrounded by glacier and near inpenetrable [sic] rain forests. Both attempts were near disasters—I nearly died the first time in, and the second I actually thought I broke my foot. Also, the site changed considerably over the past summer—the gold formerly sat in a tiny box-basin, on the very top of a formidable mountain—now, the basin has collapsed sending the material down the mountain's face. Too bad, because the first time in I hit the mountain from the bottom, the second time I hit it on top and from behind. At least I now know what to expect. It figures the next time in I will simply helicopter to the bottom of the mountain, and using metal detectors trace the mineral up; although my primary concern will be locating the mineral.

I am not concerned that anyone will find the site for some time—the area is abject wilderness, and it will take years for the placer (now released from the basin) to wash several miles to more accessible reaches. The site is in the U.S. Forest land, and is up to date unstaked.

In writing you last year, I was simply feeling out the possibilities of 'going in' with a big concern. As you can imagine, more than anything else, I simply want to find the gold—and then move onto bigger and better things. I can see that it is improbable that I will find anyone other than private parties who will grubstake me. The big companies expect me to come across with all the information—letting *them locate the mineral and stake the site. How can they expect that kind of cooperation?*

Oh well; thanks for your consideration. For now, good luck, and all best wishes, Mr. Lamarre.

Yours truly,

LeRoy Shipman Jr[1]

I don't remember what became of his property, but this interaction certainly was memorable. I was so impressed with his enthusiasm and optimism that had I not been employed by a major mining company, I might have considered joining him in his exploration efforts.

On the first of May, I flew to Albuquerque and drove to Taos where Dave Jones and I started to hike in the snow to the Gallina Peak molybdenum prospect almost due west of Taos, again in the Sangre de Cristo Mountains. Even though this property was not as high as our nearby Wheeler Peak prospect, we had to give up halfway there because the snow was just too deep. Dave had discovered molybdenite in outcrop near Gallina Peak earlier in the year and had staked twenty-nine mining claims there.

While looking at the nearby La Virgen massive sulfide prospect the next day, it started to snow heavily, so we decided to give up for now, and I flew home. A few days later, I was back in New Mexico, this time in Santa Fe to meet with the Sierra Club and the New Mexico Wilderness Coalition to discuss our Wheeler Peak property and the Wheeler Peak Wilderness Area. Of course these organizations were strongly opposed to any plans we might have for disturbing any ground near Wheeler Peak; however, it turned out to be a productive meeting anyway in that issues and positions were clarified. We could continue to work on our claims.

---

1. Permission granted on March 31, 2015 by Frederick L. Shipman, sole remaining heir of LeRoy Shipman, Jr., to print this letter.

Copper and molybdenum commodity prices had been declining for a number of months, so in late spring I began to redirect the southwest district's efforts away from exploration for these two metals. We began to focus more heavily on gold and silver, the so-called precious metals; their prices were up. To initiate this strategic redirection, I drove to the Phoenix airport where I picked up a Noranda mining engineer from Denver, and we continued on to Yuma, Arizona, to look at the old Yuma Silver (and gold) mine north of there. This gave us an initial idea of what the geology of precious-metals deposits in Arizona looked like.

Yuma was cool on this day, just under one hundred degrees, and the Sonoran Desert countryside was its usual barren and rugged self. The Sonoran Desert is the largest and hottest desert in North America and is mostly just wide-open space in southern Arizona, southwest New Mexico, and the very southeastern-most part of California. The barren landscape stretches for miles to the horizon as your vision pans from north to south and east to west.

Unlike the Sahara Desert, this desert does not have massive sand dunes; instead, it is covered by drab rock, thin sandy soil, and sparse vegetation. Typical of the Sonoran Desert were many types of cacti: prickly pear, barrel, cholla, yucca, agave, and, of course, the signature saguaro cactus with its many arms. Almost unbelievably, some of the saguaros are up to 250 years old. The saguaro cactus grows almost exclusively in Arizona; only a few survive across the Colorado River in California and to the south in northern Mexico. If you've ever seen a western movie supposedly depicting some place in New Mexico or Texas, and you see saguaros in the scenery, you'll know it was filmed in Arizona, probably at the Old Tucson movie set—Old Tucson Studios—on Tucson's west side.

In addition to cacti, the Sonoran Desert hosts ocotillo bushes and mesquite and palo verde trees. Palo verde means "green stick" in Spanish, and the bark of these trees has a pretty greenish-black color. In the spring, these trees sport beautiful lemon-yellow flowers. Filling in between the trees and cacti are shrubs: acacia, creosote bush, greasewood, and jojoba. If you're lucky as I was one day, you might even see a gila monster in amongst the cacti.

The desert environment grew on me, and I learned to appreciate its unique, stark beauty. In March, when all the cacti are in bloom, it is especially pretty. For a place that during most of the year is draped in browns and grays, the cactus flowers were a riot of bright, vivid colors in the spring. And as long as I had appropriate field gear, it was manageable working in the desert heat even though temperatures sometimes went above a scorching 115 degrees. My gear normally included a hat and long-sleeved shirt, a field vest, my sturdy field boots, lots of sunscreen, lots of water, and a comb, of course. And I never went anywhere without my trusty rock hammer.

In early June, I took a four-day trip to the Las Cuevas fluorspar (fluorite) mine in San Luis Potosi, Mexico, which was operated by Compañia Minera Las Cuevas S.A. Noranda Mines, Ltd. owned 49 percent of this mine, the largest fluorspar mine in the world at the time, and it still is. In the early 1980s, the Mexican government was considering expropriating mining properties that were a potential source of uranium for nuclear power. Because small amounts of uranium are known to accompany fluorite, Noranda Mines was extremely concerned that they might lose their Las Cuevas mine to government expropriation. Because Las Cuevas was one of the highest-grade fluorspar mines in operation, this would have been a big loss to the company; some ore contained as much as 86 percent CaF_2.

Noranda needed a radiometric survey conducted at the mine to determine if any uranium was present, and I was designated to conduct that survey. As a result, I flew to Mexico City and spent the night at a Holiday Inn. My radiometric survey gear traveled separately from Tucson by car driven by an individual who knew how to get things across the border—I didn't ask any questions. A driver picked me up the next morning, and we nearly flew (by car!) all the way to San Luis Potosi, several hours' drive north. It seemed that the rules of the road called for bullying your way through and not giving an inch to fellow drivers. This was hazardous duty.

We safely arrived at San Luis Potosi where I spent two days underground and on the surface conducting the radiometric survey, but I detected no uranium. Because I had studied the Salado Mountains fluorspar property in New Mexico for my master's thesis, I was excited to see a fluorspar mine in production, and I enjoyed the experience. How exciting it was to watch the miners drive their ore-haulage trucks up out of the mine, at a fast rate of speed, of course. Erasmo Teran, a geologist for Compañia Minera Las Cuevas S.A., was my gracious host. I then flew home, wrote a report, and that was the last I ever heard of the issue. Noranda Mines did not lose its property, and Las Cuevas remains one of the most productive mines in Mexico today.

A few days later, I was off to look for molybdenum again in northern New Mexico. This time Dave Jones and I were able to hike all the way into the Gallina Peak property—the snow was too deep last time—but I nearly walked my legs off. The exercise was good though, because it kept me fit. All field geologists are thin, during the field season at least, because we get so much exercise. In fact, someone was once heard commenting, "Those geologists are suffering from gone-ass"; they had lost so much weight from their backsides. My wife says sheepmen suffer from that same ailment. She can say that with authority since her father had been a sheepman in Wyoming.

Much of an exploration geologist's fieldwork could be considered a daily grind: hiking long distances, often in steep and difficult terrain and sometimes in inclement weather; collecting rock samples for later analysis and hauling them around in our backpacks; examining rocks with a hand lens to determine what they are; pinpointing one's location on topographic maps; and recording on maps or aerial photographs the rock types and other geologic features encountered. This is all part of being a geologic detective—trying to learn what story the rocks have to tell. And they always do have a story. You could think of rocks as kinds of time capsules that carry signatures of past great events that shaped our planet Earth. What kept me going was the excitement of knowing that around the next ridge or in the next canyon might be the clue that unravels a geologic mystery. That would define a good day.

While in New Mexico, Dave took me to another prospect where he had discovered molybdenite in an outcrop on some unclaimed ground. Good for Dave; he was being successful! To get to this prospect—called South Fork—required hiking a steep trail that traversed 2,500 feet of relief (elevation change) in only two and one-half miles. Once we arrived there and further evaluated the geology, we decided the prospect deserved more work. Dave stayed there to stake several mining claims covering the best-looking real estate, and I went back to Tucson. This prospect was also in the Sangre de Cristo Mountains, only about four and a half miles from our Gallina Peak property.

Later in June, I returned to northern New Mexico, this time with Geoff Snow to show him the prospects we had found. While looking at the South Fork and Gallina Peak molybdenum properties, we viewed the latter in a hailstorm. By this time, the Sangre de Cristo Mountains had shown they contained many occurrences of molybdenite, and Geoff agreed that we were justified in devoting a lot of attention to the area. Maybe one of these prospects would pay off for us. We knew we were in potentially favorable molybdenum country since Molycorp, Inc.'s Questa molybdenum mine was located only ten miles away to the north, also in the Sangre de Cristos. The geology at Questa served as our exploration model for the reconnaissance we were conducting.

Upon arriving at our office the day after the July 4 holiday, we were met with a scary scene—a hole had been blown through the roof of Noranda's single-story Tucson office building, specifically, right over my office. Debris was everywhere—on the floor, on bookshelves, and on my desk. That night, we were bombed again! We had known that the patrons of the bar next door often became rowdy, and over time we had kindly asked them to be less so. They apparently did not appreciate our input, and in their exuberance celebrating the Fourth of July, they decided to apply their skills as explosives experts by tossing sticks of dynamite onto the roof—they were miners, after all. We worked with detectives from the US Bureau of Alcohol, Tobacco, and Firearms to find the perpetrators, but no one was ever arrested. More hazardous duty! No one told me of this possible hazard while I was in geology school!

By mid-July, we were focusing much of our attention on the previously ignored—by Noranda, that is—gold-silver potential of southern California's Mojave Desert. A modern-day gold rush was underway there, and for the next two years we were part of it. Bob Page and Jim Norris (another Tucson-office geologist) were looking at numerous old gold prospects in California's New York Mountains, and when I joined them in the field for an update, they showed me the favorable geology that existed around some old mines: Vanderbilt Gold, Oro Belle, and Morning Star. Although not profitable to mine by the old-timers, if sufficient low-grade gold were present, we could possibly make a mine out of one or more of them.

Bob and Jim continued to have success locating favorable properties, so in late September I returned to the Mojave Desert to look at more gold prospects with them, this time in the Whipple Mountains and around Mountain Pass. This country was fascinatingly stark and barren except for its Joshua trees, the signature plant of the Mojave Desert. Because this desert is extremely desolate, you can go a long time without seeing any signs of humanity. During most days in the field, we never saw another human being.

Returning from California on one of these trips, Bob dropped me off at the Phoenix airport so I could fly to Albuquerque (again!). By evening, I had driven to Taos where I spent a few days assisting Dave Jones in geologically mapping the South Fork molybdenum property. How fun and exciting it was to scamper through the pine trees in search of outcrops.

For a district managers' meeting in Denver in late August, for some reason I decided to drive there rather than fly. My route through northeastern Arizona took me through Petrified Forest

National Park, a fabulous sight within the Painted Desert where well-preserved silicified logs are unusually abundant. Not really a forest, this is a collection of large tree trunks that were rafted together by flooded streams, buried by stream sediments, and subsequently impregnated with water-borne silica that preserved the ancient logs in a rainbow of colors; they became "petrified." The logs are almost all lying horizontally in what was once a 225-million-year-old tropical floodplain.

Some geologists suggest that these trees may have succumbed to a fate similar to the destruction of trees on the slopes of Mount St. Helens when it erupted in 1980. What a stupendous example this was of ancient geologic miracle-work conducted on the Earth's surface for all of us to see! Also of interest not far from the Petrified Forest are unusual wells that extract helium from the Permian (251–299 million years old) Coconino Sandstone; I didn't get to see them, however.

Jim and Ellie Leavitt were a husband and wife geologic team working out of Noranda's Tucson office. Recent graduates from the University of Oregon, they were especially memorable by the fact that he was as tall as she was short. They were also extremely enthusiastic. While prospecting in western Arizona, they had discovered the Socorro Reef gold prospect near Salome, just northeast of Quartzsite. I examined the property with them in August, and over the next few months we spent many days mapping and sampling the area to determine if there was potential for ore.

Returning from a visit there once, Jim and I were stopped for an hour or so by a flash flood in an arroyo that closed the highway. We just had to wait it out. At another time, I was prevented from getting to a different prospect by rushing waters of the Gila River. That was a surprise since that river seldom has any water in it at all. In all the time I spent in the southwest, flash floods were normally not a problem for me, although many times I did see evidence of past flooding.

In mid-September, we in Noranda's Southwest district office hosted a meeting of Noranda Exploration's Board of Directors, many of whom hailed from Toronto. As part of the meeting, we toured the surface operations of Noranda's Lakeshore porphyry copper mine near Casa Grande, Arizona. I had never paid much attention to this mine before. From Casa Grande, we then flew to Las Vegas and toured some of the gold properties Bob and Jim had evaluated earlier in the year, such as Vanderbilt Gold and Morning Star. The Toronto folks, however, did not display as much interest in these properties as we did.

Then it was back to northern New Mexico for me. If I've counted correctly, this was my seventh trip to that state in 1982. The objective this time was to look at gold prospects in Precambrian rocks at Gold Hill near Questa and at the Pecos Mine a few miles east of Santa Fe. We had recently started a New Mexico reconnaissance program exploring for precious metals in Precambrian rocks; led by DeWitt Daggett, another young Tucson-office geologist, the program had already identified these two properties. During the fall, DeWitt geologically mapped and sampled these prospects.

In early November, I again traveled to Denver, this time to attend a DREGS (Denver Region Exploration Geologists' Society) Symposium—more lifelong learning. While in Denver, I talked with geologists from Molycorp, Inc., about our South Fork and Gallina Peak molybdenum properties in New Mexico. Since Molycorp was the operator of the nearby open-pit molybdenum mine at Questa, New Mexico, we thought they might be interested in joint venturing with us as we explored our two properties. They declined. Did they know something about the geology there that we did not?

At year's end, Bob Page and I drove to Needles, California, and looked at gold properties north of there in the Castle Mountains and Resting Spring Range. Returning from there through Shoshone, California, we saw in a road-cut a textbook example of shiny, black volcanic glass called obsidian. The glass, in a solid layer several feet thick, angled up the excavated hillside. This volcanic glass formed when lava cooled so quickly that it didn't have time to form crystals. The glass can be broken into razor-sharp shards, so it was common for Indians to fashion arrowheads from it.

This had been a difficult year for the Tucson office staff. Although they were demonstrating success in identifying prospects of interest (they were so successful that it kept me on the road trying to keep up with them), they were working under high-stress conditions. Perhaps they should have received hazardous-duty pay because of the office bombing and battling snowstorms. And to top it off, they had to put up with a 25 percent cut in next year's budget. What would this mean for the coming year?

14. TRESPASS AT YOUR OWN RISK

At the beginning of 1983, I was on the road again; after all, that is what we exploration geologists do. My first trip was to Denver to meet with Conoco geologists and review their geologic data from the Ortiz gold-skarn property northeast of Albuquerque. (Skarn is a variably-colored green or red metamorphic rock formed where igneous intrusions come in contact with calcareous rocks like limestone.) Based on an earlier review of published literature, I was attracted to this property on which Conoco had conducted a small amount of work, including drilling.

My data evaluation in Conoco's office provided further encouragement, so in early February I went to Albuquerque to examine Conoco's drill core that was stored there. Still encouraged, the next day Dave Jones and I drove north to examine the property on the ground; we encountered snow drifts up to four feet high. What did we expect? It was February, after all! In spite of the snow, we reached the property and spent a good day in the field. Nothing beats actually looking at the rocks. But for whatever reason, we never did make a deal with Conoco or conduct any detailed exploration work there; the rocks must not have been speaking favorably to us.

Early in the year, we initiated a gold exploration program in the southwest United States based on the Carlin, Nevada, gold model: that is, disseminated, low-grade gold deposits in sedimentary rocks. I put Ellie Leavitt in charge of that program, and as a result of her diligent literature search, we became intrigued by the gold potential of the Apache Hills in the "boot heel" of southwest New Mexico. Within a stone's throw of old Mexico in the Basin and Range physiographic province, the Apache Hills form one of many linear mountain ranges separated by broad basins.

One day Jim and Ellie Leavitt, DeWitt Daggett, and I drove to New Mexico and had our first look at the area in the field. We found some prospects that might be promising, and one of us observed, "This part of New Mexico sure is thinly populated and poorly watered." I don't think we saw another soul the whole time we were there, and not much vegetation either. But I enjoyed it.

From there, I drove to El Paso to meet with geologist Noel McAnulty, Jr., to discuss New Mexico's Carlin-type gold potential. A geologic consultant who had broad familiarity with New Mexico's geology, Noel and I had worked together back in my Midwest Oil days, and I respected his expertise and opinion. After a fruitful and enjoyable discussion with Noel, I met with my team to talk with them about what I had learned from him; as we had thought, this part of New Mexico had potential. So we brainstormed and eagerly decided to acquire some claims that had already been staked by another prospector on one of the Apache Hills properties we had seen.

But first, I had to go see what Bob Page and Jim Norris were so excited about. They had identified an attractive property at the southern end of the Sierras a few miles northwest of the town of Mojave. This required us to get up to speed on Sierra Nevada geology. This mountain range is comprised primarily of granite plutons (bodies of igneous rock up to several miles across) that you can think of as being fossilized magma chambers. These plutons formed five to ten miles below the Earth's surface as magma cooled and crystallized into rock 80–120 million years ago. As cooling progressed, the plutons fused to form a huge body of igneous rock called a batholith; you can visualize it as resembling an assemblage of large marshmallows. In the case of the Sierra Nevada, the batholith is over four hundred miles long. Your grandmother's homemade rolls sitting in a bread pan is another good, small-scale proxy for a batholith.

Then, just five million years ago during the Basin and Range orogeny (a period of mountain building), the eastern margin of the Sierra Nevada batholith started to rise, and erosion stripped off the overlying rocks, exposing its granitic core. I've always found it intriguing and ironic that the rocks of the Sierra Nevada are modestly old (80–120 million years), but the topographic feature that constitutes that spectacular mountain range is young (five million years).

On my first visit to Bob's gold prospect, called the Loraine project, I learned that he has a good sense of humor—he had named the claims he staked the "Quiche claims." Never mind that "Loraine" was not spelled correctly! After mapping and sampling here, Bob began drilling the prospect later in the year.

The Loraine property is on the west side of the Mojave Desert, and to reach it from Arizona requires a drive completely across the desert from Needles, California, on the Colorado River. The Mojave Desert of Southern California is as barren as the Sonoran Desert, if not more so. The only significant difference I could see is the absence of saguaro cacti, but in some places the Mojave Desert has "forests" of fuzzy-looking Joshua trees. These strange-looking trees define the Mojave Desert and are found nowhere else. Temperatures in this desert often exceed a blistering 115 degrees in late July and August, and because it lies in the rain shadow of the Sierras, this desert receives even less precipitation than the Sonoran Desert—some areas average only three inches of rain a year. The small hamlet of Bagdad (which no longer exists) once went 767 days without measureable precipitation.

To get to the Loraine project from Tucson requires traveling from Needles to Barstow, California, about two-thirds of the way across the Mojave Desert. The route is 140 miles of mostly nothing except creosote bush. I suggest not traveling this route if your vehicle is not in good shape; it's not a good place to become stranded. The route passes through Ludlow and Hackberry Springs, but they're just small places wanting to be towns.

The east-west Interstate 40 across the desert follows much of the former route of the iconic Route 66 on its way from Chicago to Los Angeles. Driving this route led me to wonder what it must have been like in the 1930s as emigrants from the Dust Bowl traveled to California on the pre-interstate highway. In those days, they didn't have a McDonald's

Drilling at the Loraine Project in the southern Sierras, 1983.

restaurant in Barstow at which to stop for refreshments like we did. In fact, the Barstow McDonald's must be the largest in the world—absolutely huge. Perhaps being situated at the intersection of Interstate 40 and Interstate 15 makes for an ideal location.

Even though the Mojave Desert covers nearly 25,000 square miles, on occasion as we hiked off-road looking at prospects, we would come across a lone travel trailer or a rickety old building, still occupied. The desert seems to attract some people who enjoy the life of a hermit—a life too reclusive for me. Some of these people are a little strange, or worse, so we made sure to give them a wide berth. Hmmm, the name Charles Manson (the cult leader and serial killer still in prison) comes to mind.

To occupy our time and break up the monotony as we drove along Interstate 40 toward the Loraine project, Bob and I often discussed what we were seeing off in the far distance. West of Barstow, we could see the largest borax mine in the world at Boron, California. This huge mine and associated processing facilities provide about half the world's supply of borax, an industrial compound that most people only know of as Boraxo, the cleaning agent. An essential component of many everyday products like detergents, cosmetics, enamel glazes, fire retardants, insecticides, and antifungal agents, borax is also an ingredient in the new vaccine Gardasil.

About twenty-three million years ago, this area was occupied by large, seasonal saline lakes that periodically evaporated, leaving behind sedimentary beds of colemanite up to 120 feet thick. The borax mineral colemanite (comprised of calcium, boron, oxygen, and water) was discovered here in 1913, and mining began in 1927. Today, these evaporite beds are mined by open-pit methods, and the operation is California's largest open-pit mine. This is so different from the days of the twenty-mule team wagons that slowly hauled the borax to processing plants; you could think of those 150-foot-long wagons as nineteenth-century semitrucks.

Immediately south of the town of Boron is Edwards Air Force Base where experimental aircraft are tested and where NASA's space shuttle landed when Cape Canaveral was weathered in. Even though it's "God-awful" hot here in the summer, when passing through during the winter months, we commonly saw snow atop the San Gabriel and San Bernardino Mountains in the distance to the south. Los Angeles lies on the far side of those mountains.

Farther west, we drove through the town of Mojave, which to me is notable for its commercial-jet aircraft graveyard. Apparently, because of the very dry climate, this is where airline companies choose to temporarily store or retire the aircraft they don't need. As we drove by, we could see passenger jets sitting out in the open belonging to most of the major airlines.

Even more fascinating, however, was the Tehachapi Loop on the railroad line even farther west. This line connects the Mojave Desert with California's lower-elevation San Joaquin Valley, and we drove by the loop on the way to the Loraine property. As the railroad winds its way up Tehachapi Pass (3,793 feet in elevation), it travels in a spiral through a tunnel such that the track passes over itself. A train with eighty-five boxcars will cross over itself as it goes around the loop. This design decreases the grade that the engines have to pull, and, amazingly, the technological marvel of this railroad loop was accomplished more than 130 years ago (1874–1876). The Chinese laborers who constructed this section of railroad can be thanked for their efforts because this is one of the busiest railroads in the world—on average, thirty-six trains travel this loop every day.

One day, instead of driving to the Loraine project, I flew into Bakersfield, California, in Kern County, where Bob Page picked me up. I knew nothing of Bakersfield other than the fact that it is a large city in California's San Joaquin Valley. I was immediately impressed with the miles and miles of vineyards we passed as we drove along. Because I knew we were not near California's famous Napa Valley, this did not seem like a high-quality grape-growing area to me. Consequently, I quipped to Bob that they must be producing "college wine"—that is, cheap wine like Ripple. My friends and I drank Ripple at Dartmouth and maintained it was the only wine that gave you a hangover while you drank it. With a screw-top bottle, it was priced just right for a poor college student. I didn't know anything about the wines produced around Bakersfield—I didn't know anything about any wine, for that matter—I was just being silly.

We also saw miles and miles of citrus crops—oranges mostly. Farther on, we passed by an oil field with its myriad of pump jacks and a refinery. Agriculture and oil are what make Kern County such a rich and productive part of California.

When back in Arizona after one visit to the Loraine project, at the invitation of the mine owners I visited the Silver Bell property northeast of Florence, Arizona. This trip was especially memorable because I got to experience something I had never done before—drive into a box canyon. Part of the access to the mine site was along North Box Canyon Road, which follows a winding, dry, creek bed in a narrow canyon where vertical cliffs hundreds of feet high are only about fifteen feet apart. Had there been a storm upstream while I was driving in the canyon, I would have been caught in a flash flood with no way out.

The mine owners were exploring the old underground mine workings at Silver Bell, so I went underground with their geologist and spent the night in their field camp. Nothing came of this visit, and after reviewing their geologic data the next day, I drove out of the canyon on my way to Yuma to look at more attractive properties: operating gold mines and prospects across the Colorado River in California.

There, Bob Page, Jim Norris, and I took a look at the old American Girl gold mine in the Cargo Muchacho Mountains where Newmont Mining Company was conducting an extensive drilling program. Their geologist was gracious enough to show us around. Newmont was having success, and six years later they put the property into production.

We then toured Consolidated Goldfields Corporation's Big Chief gold mine in the nearby Chocolate Mountains, which Goldfields had put into production just the year before. This was turning into a good trip because the geology at these mines was fascinating, and the temperature was not very hot—only eighty-five degrees. We then went to Glamis Gold's operating Picacho gold mine a few miles from the Colorado River at the base of Picacho Peak. The temperature was a little hotter there—we collected samples in the 108-degree heat. These three properties clearly signified that California's Cargo Muchacho Mountains and Chocolate Mountains, located only a dozen or so miles northwest of Yuma, Arizona, were good places for gold exploration. It's always wise to go where the gold is.

These mines exploited deposits where small concentrations of gold were disseminated in large volumes of highly fractured rock. Concentrations of only 0.05 ounces of gold per ton of rock mined were typical. Imagine it, there was far less than an ounce of gold in every ton of rock mined. These exceedingly small gold concentrations were economical to mine for two reasons: first, tens of millions of tons of rock could be efficiently mass-mined by large pieces of mining equipment, and second, the advent of cyanide heap leaching was an efficient and inexpensive method of recovering the gold. Using this method, crushed gold-bearing rock is heaped into huge mounds onto which a dilute cyanide solution is sprayed. As the solution percolates down through the pile, gold is dissolved by the cyanide and carried in solution. This "pregnant" solution is captured at the base of the mound, and the gold is then precipitated out of solution. This would have been just waste rock to the old-time miners. How things change.

The granites, gneisses, and schists of this part of California had been cut by nearly flat-lying faults that provided migration pathways for gold-bearing hydrothermal fluids; within the highly fractured and sheared zones, gold had precipitated. The three fracture-controlled deposits we looked at became our geologic model for gold exploration in southeastern California.

Since last fall's geologic mapping of the Socorro Reef property near Quartzsite showed that it had potential, we got busy during the spring negotiating lease terms with the owner, securing drilling permits from the State of Arizona, and arranging for a drilling contractor. By late March, we were nearly ready to start drilling, so Jim Leavitt and I drove there to layout drill sites. By early April, reverse circulation drilling was underway. This type of drilling grinds up the rock and spits it out at the surface as small chips, called cuttings. As in core drilling, we then geologically logged the chips and collected some for assaying.

At one point, we had geologists from other Noranda offices visit the property to provide their geologic insights. Temperatures were well above one hundred degrees, and we looked pretty bedraggled as we stood under our tent examining drilling samples. Since the property was not too far from Tucson, I took Javier Contreras, our office draftsman, and Debi Busser, our secretary, to the property one day so they could see what a drilling operation looked like. I think they decided they preferred office work.

During the first week of April, I was off to Denver again for another district managers' meeting; this one included a daylong seminar on how to conduct successful property negotiations. Even though Noranda had a full-time landman who negotiated deals for us, we field geologists usually initiated discussions with property owners, so this was a valuable course.

By the end of April, I was back in southwest New Mexico because the geology in that part of the world was looking pretty attractive. Ellie's research into the geologic environment of gold occurrences there was paying off; it must have been the rocks were speaking to her. In addition to prospects in the Apache Hills, a newly discovered prospect in the nearby Animas Hills, called Winkler Anticline, looked even better. I was somewhat familiar with the Winkler Anticline because I had been there years ago when I was working for Midwest Oil in the Salado Mountains.

We ended up optioning (leasing) some mining claims at the Winkler Anticline and began mapping the geology and collecting rock samples for analysis. At the same time, we looked at the Klondike jasperoid-gold prospect and the Zebra prospect, both nearby. Those days in the field were wonderful because of attractive geology and great weather.

On one of my driving trips to southwest New Mexico, my fellow geologists and I had fun playing around on copper anodes stacked neatly on a flatbed railroad car at a rail siding near the highway. Anode copper is nearly pure copper metal (99.6 percent pure)

The author (on the left) and Jim Norris enjoying the
discovery of copper anodes on a railroad car, 1983.

that is produced by smelting (melting) concentrated copper-bearing minerals mined from open pits or underground mines. Smelters produce molten copper that is poured into molds about three feet by three feet by two inches thick. When the liquid copper cools, it hardens into solid copper having this characteristic shape with handles or lugs on which to hang them during the subsequent refining process. They're called anodes because they serve as the positive electrode (the anode) during electrolytic refining at refineries, which further purifies the copper.

The copper in these anodes probably ended up as copper wire or pots and pans. Weighing 800–900 pounds each, it's a good thing we didn't drop any of these anodes onto our toes. These copper anodes are the product of what many geologists had spent much of their careers exploring for, and we couldn't resist having our picture taken next to them. Thank goodness the highway patrol didn't come along while we trespassed.

However, in mid-June I *was* busted, but it was in West Texas! Intrigued by the potential for discovering gold in Proterozoic rocks there, Vic Chevillon from Noranda's Missoula office and I went exploring in the Van Horn area southeast of El Paso where we experienced the long arm of the law. As we drove along, we saw an outcrop in a field a little way off the highway, so we decided to stop, climb over the fence, and take a look. No harm in that, is there? Well, yes there is! Unlike in other states, all land in Texas is privately owned, and the ranch owner didn't take kindly to our beating on the rocks on his *private* property.

Upon returning to our vehicle after a short traverse through the rancher's barren field, we were met at the fence by the rancher's daughter and her trusty shotgun; it was pointed right at me from just a few feet away. Clad in a dirty, white T-shirt and raggedy jeans, her body language meant business—she was wound tighter than a spring. "What do you think you're doing? This is private property," she said with an intense sneer on her less than attractive face surrounded by long, stringy hair. It was obvious this sturdy, rough-looking blonde was not to be trifled with.

Her older brother was called, the constable was called, the sheriff was called, and then her father arrived. He mumbled something about his land, paying taxes, and "Who do you think you are?" and the Constable informed us in no uncertain terms that we had trespassed on the rancher's land and violated his privacy. The sheriff arrived in a cloud of dust and told us, "This old boy doesn't want anybody on his ground. We'll let the judge decide. You can either follow me or …" Whereupon we were escorted with lights flashing the entire twenty miles to the Hudspeth County courthouse and jail in Sierra Blanca, Texas.

On our escorted drive to town, Vic and I tried to guess how much it would cost to get out of this mess, and we wondered if we would have enough cash in our pockets to pay the fine. This was late on a Friday afternoon, and we didn't think Don Alberts, the office manager of Noranda's headquarters office in Denver, would have time to wire money to us. We certainly were not looking forward to being guests for the weekend at the Hudspeth County jail.

The decider of our fate, Judge Doyle Ziler, was attired in a rumpled, gray, cotton suit that matched the worn paint on the concrete floor of his office. He was a genteel, scholarly enforcer of the law, and justice was expeditiously dispatched. "Guilty!" After duly admonishing us, he fined us ninety-six dollars each, which we were able to cover. Whew! You can bet that went on my expense account! After saying our "sorrys" and "We'll never do it again," we hightailed it out of the state and never conducted exploration in Texas again! Ever since, Texas has not been my favorite state. Someone once said that bad decisions make for good stories, and that was certainly supported here.

Vic and I received a prestigious company award for surviving this unfortunate incident—the coveted "Noranda AWSHIT Award", one of my most precious possessions. It states:

> *"Everyone knows that 1,000 ATTABOYS qualifies one to be a leader of men, and that one AWSHIT Award returns the recipient to square one. It grieves me no little, therefore, to be faced with the need to tender this dubious award. I would not, indeed could not, perform this reprehensible task were not the facts so clear, the deed so onerous, the culpability so apparent. It is with great pain, much distress, and considerable chagrin that I hereby award, not one (1), but two (2) AWSHITS, one to Albert L. Lamarre and another to C. Victor Chevillon for getting CAUGHT (ugh). Shame on you both! —G. Snow"*

This event would come back to haunt me, however. Years later as I was filling out paperwork to obtain a US Department of Energy security clearance, which would allow me to work at Lawrence Livermore National Laboratory, one question asked about any arrests. Oh dear! I puzzled over this for fear that admitting so would prevent my getting the job, but withholding the information could be damaging later if it were ever found out. My conscience ruled, and I did fess up; it apparently had no adverse effect on my being hired, and I ultimately received a high-level security clearance to boot.

Three days after the rifle-pointing event, I flew back to El Paso and rented a car to quickly drive to New Mexico—a more receptive state for my efforts—to look at the White Oaks property near Carrizozo. This town is merely a collection of buildings near some isolated hills southeast of Albuquerque. From those hills, the flat Great Plains of eastern New Mexico are visible for miles as they stretch out and fade into Texas. A good place for an alien spacecraft to land, methinks. Pop culture of the 1970s said that's exactly what had happened when, as UFO advocates alleged, extraterrestrial debris—including alien corpses—was found in and around a flying object that crashed near Roswell, New Mexico, in 1947. Conspiracy theorists maintained that bodies were recovered from the alien spaceship and the US military was involved in a cover-up. The alleged event even elicited congressional inquiries and an investigation by the General Accounting Office. Since then, Roswell has been synonymous with unidentified flying objects (UFOs). The town even holds an annual UFO Festival.

After many more driving trips to Winkler Anticline to finish the geologic mapping and sampling, we decided the prospect warranted drilling. Once approval was received from the US Bureau of Land Management of our Plan of Operations, Ellie, Jim, and I picked out drill sites, contracted with a local bulldozer operator to build drill roads, and supervised a contractor who conducted a geophysical survey for us. All of this took time, and it wasn't until the spring of 1984 that drilling actually began. I even had to make one trip to Denver to conduct further negotiations with the owner of the claims we had optioned.

On one of my many driving trips to New Mexico, I had two flat tires in one day. The second one left me stranded in the middle of nowhere, so I had to walk to the nearest settlement, while rolling my flat tire, to get help. Of course we had no cell phones or CB radios in those days. This caper taught me to always carry more than just one spare tire. Once I got the tire fixed, it cost me twenty dollars to retrieve it and get a ride back to my truck. After that problem was taken care of, I began driving to northern New Mexico on Interstate 25 when I got a speeding ticket north of Truth or Consequences! That just wasn't my day.

Later on, I drove to Salome, Arizona, to check on the drilling Jim Leavitt was doing at Socorro Reef. While there, I picked up 250 feet of drilling samples (we collected a bag full of drill cuttings from every ten feet of drilling) to deliver to Skyline Labs in Tucson for gold analysis. But alas, we found out a few weeks later there wasn't much gold in the samples, and we dropped the property in late September.

Right after that, I drove almost twelve hours from Tucson to Taos where the next day I backpacked with my tent, camp stove, sleeping bag, food, and geologic paraphernalia up to Gallina Peak in the Taos National Forest. I camped by myself for the next four days at about ten thousand feet while geologically mapping this molybdenum property. The weather was mostly gorgeous at this high-mountain location about ten miles north of Taos. What a peaceful setting.

Although it had been quite a while since I had mapped a property in detail by myself, I thoroughly enjoyed it. I found quartz-pyrite-molybdenite veins cutting various kinds of igneous and metamorphic rocks, but unfortunately the geology did not fit our geologic model. Although the intrusive rocks at Gallina Peak were part of the Questa igneous system that produced the Questa molybdenum deposit, they were older than Questa's ore-producing rocks and had no potential for yielding ore. My geologic mapping showed that these rocks did not meet the requirements of our exploration model. As you can see, exploration is not a haphazard or random process but instead relies on having a geologic model to guide your exploration efforts. Consequently, we did no more work at Gallina Peak and dropped the mining claims.

At the end of September, the entire Tucson office geologic staff flew to Reno for Noranda's annual meeting of the Witchers and Dowsers Society. The focus this year was the geology and ore deposits of the Basin and Range Physiographic Province. This province, centered in Nevada, is an immense region of north-south-trending, faulted mountain ranges that are separated by broad, deep, sediment-filled valleys or basins. As viewed from satellite photos, the province presents a corrugated landscape that sprawls from southeast Oregon, through Nevada, western Utah, eastern California, Arizona, and New Mexico to west Texas. About five hundred mountain ranges comprise the province, and it has no counterpart anywhere else in the United States. The province is bounded on the northeast by the Teton Mountains in Wyoming and on the west by the Sierra Nevada.

This fascinating province was created starting approximately twenty million years ago when the Earth's crust began to stretch east to west and cracked north-south into dozens of crustal blocks; some rose to become mountain ranges, and some fell to become basins. You can think of these as stretch marks in the Earth's crust. Although difficult to visualize, as a result of this crustal spreading, the distance between Reno and Salt Lake City today is about twice what it was before spreading began. Believe it or not, because Basin and Range faulting is ongoing, each year Sacramento moves about one-half inch farther away from Salt Lake City.

The Witchers and Dowsers Society field trip took us to the Great Basin, the part of the Basin and Range Province in Nevada where surface waters drain inward and do not reach any ocean. Water that falls in the Great Basin stays in the Great Basin. Most of what we saw there were vast open spaces and broad basins paved with sagebrush and other hardy plants. The mountains are islands of green whose cooler temperatures support forests of juniper, pinion, and pine trees. The basins are filled with thousands of feet of sediment, and few outcrops are present, so we concentrated on looking at the mountain slopes.

After starting the trip in Reno, we spent one night at the June Lake Lodge near the ski slopes of Mammoth Mountain, California, on the western margin of the Great Basin. This stop was appropriate because June Lake lies at the foot of the Sierra Nevada, the mountain range whose rain shadow produces the dry and bleak terrain of Nevada's Great Basin. High above the lake, we saw cirques, arêtes, and tarns, all features carved out by the mighty force of glaciers. Again we have geology to thank for the stunning scenery.

The Sierra Nevada hosts twelve peaks more than fourteen thousand feet high, and their elevations were attained in the relatively recent geologic past. In the grand scheme of things, these mountains are mere babies. Starting five million years ago, *más o menos*, the Sierra was uplifted along a ten-mile-wide band of faults called the Sierra Nevada Frontal Fault System. Through many, many discrete movements measured in inches or a few feet along individual faults, the once-buried granitic rocks rose to austere heights where they fought the erosive forces of wind, water, and glaciers. Nature abhors high elevations and is always wearing down the rocks at those elevations. The Sierra Nevada Frontal Fault System is the westernmost boundary of the Basin and Range Province.

This Witchers and Dowsers Society trip introduced me to a fascinating geologic setting that was new to me, and it turned out to be another educational and wonderful experience. But it was time to get back to work in the southwest district, so back to Tucson we went.

Early in October, I was looking at prospects around Twentynine Palms, California, with Bob Page and Jim Norris when I began to feel weak. I spent the drive back to Tucson sleeping in the backseat of Bob's Bronco. When back in Tucson, I ended up spending four days in the hospital. The doctors determined that I had lost a significant amount of blood, but it took them four days to identify a bleeding ulcer. *Now how did I get that?* I wondered.

In early November, I flew again to Reno, this time to accompany Larry Lackey and Greg Cox, both from Noranda's Reno office, to look at properties around Austin and Ely (pronounced E-lee), Nevada, and in western Utah. To reach these properties, we traveled "The Loneliest Road in America," although it wasn't called that at the time. However, I do agree with the name. In their July 1986 issue, *Life* magazine described Nevada's part of US Highway 50 from Fernley in the west to Ely in the east as "The Loneliest Road in America." The article said there were no attractions or points of interest along the 287-mile stretch of road and recommended that drivers have "survival skills" to travel that route. In that article, a AAA representative remarked, "It's totally empty. We don't recommend it. We warn all motorists not to drive there unless they're confident of their survival skills."

Larry, Greg, and I traveled that loneliest road in America, and I can attest to the fact that it is lonely. However, it does have attractions, at least for a geologist. Basin, range, basin, range was the rhythm of the route as we cut across the grain of the province. Not a good place to break down in your vehicle; help could be a long time coming. When the sign says, "Next Gas 167 Miles," you'd better believe it.

Much of the highway follows the old Pony Express Trail and the route of the Overland Stage through Fallon, Austin, Eureka, and Ely, Nevada. These old mining towns are nearly ghost towns now, each situated on a mountain range and reminding us of what life must have been like in the 1800s. Janet and I got to enjoy this scenery years later when we drove the route on our Harley-Davidson motorcycle. We especially enjoyed seeing the Shoe Tree east of Fallon. Here, many earlier passersby had removed their shoes, tied the shoe laces together, then tossed their shoes over limbs of a tree beside the road. There were hundreds of them. Although not an approved Nevada landmark, it certainly was colorful!

When back in the southwest in late November, I accompanied Jim Norris to Parker, Arizona; Needles, California; and Las Vegas,

Nevada, where we looked at several more gold prospects along the Colorado River. Names that I remember are Coliseum, Savahia Peak, Morning Star, and Rattlesnake.

Things were looking good: we were drilling properties and finding good-looking prospects in three states.

15. THE ADVENTURES
COME TO AN END

Okay, our budget for the year 1984 had been cut by 25 percent. But that didn't stop us from searching for that elusive ore deposit. In January, I expanded my exploration repertoire once more by learning about a geologic setting that was totally foreign to me: Mississippi Valley-type lead-zinc deposits of the country's midcontinent. This was another continuous learning experience, courtesy of a district managers' meeting. I flew to Omaha, Nebraska, then to St. Louis where I rented a car and drove to Rolla, Missouri, to join the other district managers. This meeting and its educational session were hosted by the staff of Noranda's midcontinent exploration office in Lebanon, Tennessee. Exploration conducted out of this office focused on the search for Mississippi Valley-type lead-zinc deposits, which are widespread in the so-called Viburnum Trend of southeast Missouri and parts of neighboring Iowa, Illinois, and Wisconsin.

The Viburnum Trend was the world's largest lead-producing district at the time. To learn about these deposits, we went underground in AMAX's Buick mine where ore is hosted by Cambrian (488–542 million years old) carbonate sedimentary rocks. These rocks are from the same geologic environment as at London, Ontario, where I went to graduate school. Although I never had the opportunity to conduct exploration in this environment, it was exciting to learn about it nevertheless.

By now all of our exploration efforts out of the Tucson office were dedicated to finding mineable deposits of gold or silver. Consequently, in early February, Jim Leavitt and I drove to Deming in southwest New Mexico to do some more gold prospecting. The route we took, Interstate 10, closely follows the old route of the Butterfield Stage. From 1857 to 1861, Butterfield stagecoaches carried people and mail from Memphis, Tennessee, to San Francisco, California, along an arduous dirt track. Thankfully, our drive was faster and more comfortable than theirs, and the next day we were out in the field looking at jasperoids in the Lake Valley–Hillsboro area. You'll remember that jasperoid is silicified carbonate rock, and in this area it sometimes contains precious metals.

What a beautiful day we enjoyed just east of the Black Range, one of the most remote parts of New Mexico. With peaks topping out in excess of ten thousand feet, we were in Apache country, famous for the great Chief Geronimo. Because of its remoteness, this area was his most important stronghold. This part of New Mexico is dotted with old mining camps that had their heyday in the late 1800s when the towns were filled with get-rich-quick schemers and miners. Most of these towns are now ghost towns or nearly so. Chloride, for instance, had a population of only seventy-one full-time residents in 1976. Silver mining throughout the west collapsed during the silver crash of 1893 when the US government selected gold as the country's monetary standard. The subsequent drop in silver prices resulted in the demise of many western towns.

Chloride is not far from the town of Hatch, the self-proclaimed Chile Capital of the World. Hatch is famous for its colorful red chili ristras that hang from building entryways throughout New Mexico in the fall. The town holds an annual fall chili festival but, unfortunately, I was never there at that time of the year.

While prospecting in the area, we drove onto the Plains of San Agustin—in the middle of nowhere—and came upon a scene right out of *Star Wars*. Here, enormous saucer-like antennae gather radio signals from far out in space. Called the Very Large Array, it is part of the United States' National Radio Astronomy Observatory, whose goal is to understand the conditions within the universe when our solar system was young. I couldn't help but wonder: *Are they broadcasting, "ET, phone home"?* There are twenty-seven of these huge, dish-shaped antennae pointing out at the heavens. Interestingly, the antennae sit on what was once a fifty-mile-long Pleistocene lake similar to Glacial Lake Missoula. What an effective use of otherwise useless terrain.

Later in February, Janet and I drove to Phoenix where we met geology student field trippers from the University of Western Ontario geology department and hiked with them up the South Mountain metamorphic core complex just south of Phoenix. (A metamorphic core complex is a domelike mountain cored by deep crustal rocks. They are common across the southwest.) Later, we all examined the volcanic stratigraphy at Superior, Arizona, east of Phoenix. I enjoyed seeing Bob Hodder again, and of course the students are always fun.

Toward the end of the month, I flew to Los Angeles to attend a meeting of the American Institute of Mining, Metallurgical, and Petroleum Engineers, which included a sightseeing tour of Hollywood, Rodeo Drive, Beverly Hills, and the La Brea Tar Pits. The tar pits were the most exciting part; imagine, fossils of 38,000-year-old saber-toothed cats, bison, and giant ground sloths encased in tar right in the middle of downtown Los Angeles.

The return flight stopped at Borrego Springs, then flew on to El Centro, California, just north of the Mexican border, where I deplaned and spent the night. El Centro sits at the southernmost expression of the San Andreas fault, probably the most widely known and intensively studied fault on Earth. This fault is famous, of course, for causing the great San Francisco earthquake of 1906 when, at a location about two miles offshore from San Francisco, the fault moved as much as twenty-eight feet horizontally, generating the magnitude 7.8 earthquake. In all, 296 miles of the Earth's crust ruptured that day along the nearly eight-hundred-mile-long fault. In Marin County north of San Francisco, I've seen a wooden fence that was offset a few feet where the San Andreas fault passes beneath it.

This is a good opportunity for me to express my strong displeasure with the popular news media's use of the term "earthquake fault" instead of just "fault." Their terminology is misleading. A fault is a planar surface along which rocks have ruptured. Excessive force, be it tensional or compressive, can build up in rocks to the point where they rupture. When the rock masses on either side of the rupture plane move instantaneously, a fault has been created. This movement generates shock waves that pass through the rocks, and we feel the vibrations as an earthquake. Said another way, earthquakes are produced when one mass of rock moves relative to another mass of rock; built-up energy in the Earth's crust is suddenly released. Earthquakes are produced *only* by movement along a fault, not by any other means. So adding the word "earthquake" to modify the word "fault" is redundant. Just say "fault."

Now back to El Centro. Adjacent to the nearby San Andreas fault is the Salton Sea, which has a fascinating history of its own. In 1905, heavy rainfall and snowmelt caused the nearby Colorado River to breach its levee, and the full extent of its flow poured into a low-lying depression. Two years passed before the flow could be stopped, and by then the Salton Sea had been created. Today, it has no outlet and is continually becoming saltier as evaporation progresses. The sea lies in the bottom of the Salton Trough with a surface-water elevation of 226 feet below sea level. As you drive along Interstate 8, a highway sign tells you when you go below sea level. How exciting!

Following my night in El Centro, fellow Noranda geologist Gordon Hughes and geochemist Pete Holland joined me to look at the old Tumco gold mine in the Cargo Muchacho Mountains of Imperial County, California, about sixteen miles west of the Colorado River from Yuma, Arizona. I had done some preliminary exploration in the Cargo Muchachos last year and determined they merited another look, this time with the assistance of Gordon and Pete.

The three of us found free gold in outcrop; that's always encouraging. Gold was discovered at Tumco in 1864, and mining from underground workings up to one thousand feet deep continued off and on until 1914. After examining the Tumco Mining District for a while, we speculated that the gold might be of volcanogenic origin (produced by undersea volcanic action like that which had produced ore at Jerome, Arizona). If our hypothesis were correct, it made these Paleozoic (251–542 million years old) rocks especially attractive as exploration targets. The rocks were speaking to us again. "We'll have to come back with a bigger crew and do some detailed work," we decided.

While in the Cargo Muchacho Mountains, we also explored the old Padre y Madre gold mine, then looked at prospects in the Chuckwalla and Chocolate Mountains, all in the southeastern corner of California. Recognized as the location of the state's first recorded mining activity, the Spanish conducted placer mining for gold in southeast California in the 1780s. The Spanish influence is reflected in the abundant Spanish place names found throughout Imperial County. It had been a productive few days spent in this obscure corner of California.

After returning to Tucson, I soon left with Bob Page to drive to Kingman, Arizona, to examine precious-metal prospects around Bullhead City on the Colorado River. On the way to Kingman, we crossed over the Hassayampa River in Wickenburg, where an official state highway department sign caught my eye: "No Fishing." I don't know when this river ever saw enough water to hold fish; however, it was nice to see a sense of humor expressed so publicly. Farther along, we passed through the town of Nothing, Arizona, and true to its name, there was nothing there.

Once we got west of Kingman, we came across a textbook example of a normal fault in a road cut on Interstate 40. (A normal fault is one in which rocks on the upper side of the fault plane move down relative to rocks on the lower side. By contrast, the San Andreas is a strike-slip fault, meaning that the rocks on one side of the fault plane move horizontally relative to rocks on the other side.) In this road cut, welded tuffs of contrasting colors were offset so that dark volcanic rocks were juxtaposed against lighter colored ones, making the location of the fault so obvious a blind man could have seen it. Where was my camera when I needed it? Even though all highway patrol departments frown upon geologists who stop on their highways to look at rocks, we *had* to stop. This road cut was presenting us a view into a world of ancient times, a world never inhabited by *Homo sapiens* but full of volcanic activity.

By the middle of March, I was back in southwest New Mexico's jasperoid province looking at the Apache Hills and Winkler Anticline properties again. I was accompanied this time by Noranda geologists Hart Baitis of the Missoula office, Eliseo Gonzalez-Urien of the Denver office, and Larry Lackey of the Reno office. They were there to provide their expertise in evaluating the two properties. Having geologists with different experience bases look at the same rocks was always helpful.

Later that month, I was off to Seattle for another district managers' meeting, this time to reallocate budgets—again! We just did that last year! The mining industry was not faring well at all because of low metals prices, and times were getting tougher for Noranda Exploration. At that meeting, all of our budgets were cut again, but everyone still had a job.

Since gold exploration in Nevada was the main focus of the mining industry's attention in the early 1980s, I needed to see for myself what these so-called Carlin-type gold deposits looked like. So I flew to Reno where Steve Zahony, Bob Page, and I headed east on Interstate 80 toward Elko, Nevada. Barely had we gotten underway before we saw a sign for Pyramid Lake, the only remnant of Glacial Lake Lahontan, which once covered most of northwestern Nevada. Toward the end of the last ice age when huge glaciers covered the Sierra Nevada to the west, melting of the glaciers filled the valleys of northern Nevada with meltwater. Although once larger than Lake Ontario, Glacial Lake Lahontan has so deteriorated by evaporation that small Pyramid Lake is all that remains of this once huge glacial lake. This geologic artifact reminded me of Glacial Lake Missoula.

Not long thereafter, we came upon a broad, featureless flat called the Humboldt Sink. This playa or dry lakebed is the endpoint for the Humboldt River; it can flow no farther in this parched environment. I use the word "river" advisably here. Having been raised in New Hampshire among the rapidly flowing Connecticut, Ammonoosuc, and Androscoggin rivers, the Humboldt did not look like any self-respecting river I had ever seen. The travelers on the mid-1800s California-bound Emigrant Trail thought the same thing. One of them called it "the most miserable river on the face of the earth. " Nevertheless, the Humboldt River is one of the most important rivers in American history because it provided access for the forty-niners to cross Nevada on their way to the California gold fields. They were never more thankful than when they reached the sink and had the Sierra Nevada in sight.

We finally reached our objective, Winnemucca, Nevada, where Steve, Bob, and I toured the Pinson gold mine and then drove on to Elko where we stayed for a couple of nights. We spent two days touring Newmont Mining Company's famous Carlin gold mine and Jerritt Canyon (Marlboro Canyon) gold mine. These operations mine sedimentary-rock-hosted gold deposits that lie in a broad north-northwest-trending belt across northern Nevada called the Carlin Trend. Some of the deposits in this gold belt contained high-grade ore, up to one ounce of gold per ton of rock. The gold in these deposits is extremely fine grained, so much so that it cannot be captured by panning. No wonder the mid-1800s emigrants did not stop here to mine; they couldn't even see the gold. Ironically and sadly, they hurriedly passed over more gold beneath their feet in Nevada than they would ever mine in California. I was impressed with the subtlety of these gold deposits.

It turns out we were one year too early to attend Elko's Annual National Cowboy Poetry Gathering. In 1985, the city started an annual celebration of life in the rural American West, and this event turned into a weeklong gathering of cowboy poets and folklorists who attract thousands of visitors each year. This celebration has made Elko famous.

Before returning to Reno, we stopped at a hot springs geothermal area at Beowawe, (pronounced bay-o-WAH-wee) Nevada, east of Battle Mountain. Beowawe is a Paiute Native-American word meaning "gate," so named for the peculiar shape of the hills close by that resemble a gate. The state of Nevada experiences high geothermal heat flow from magma lying at shallow depths, and the hottest crustal area is around Battle Mountain, where temperatures in the Earth's crust are twice the continental average; reservoir temperatures can reach 400°F. Thus, Nevada has good potential for geothermal energy, and exploration was underway there for this energy resource.

The most conspicuous feature of the Beowawe Geysers area is an enormous, symmetrical sinter (crusty material) terrace—formed by evaporation of hot-spring water—standing about 250 feet above the valley floor. I was impressed with the top of the terrace, which measures one hundred feet wide by twenty-five hundred feet long, because it is remarkably smooth and level. The hot springs, geysers, and fumaroles there have temperatures up to 200°F, and in 1932, several geysers were reported to have erupted to heights of more than three feet. We saw exciting rocks on this trip, and the weather treated us well with beautiful, cool days spent beneath the snowcapped mountains.

In late March, I received a visa from the Mexican Consulate in San Francisco that enabled me to enter Mexico to conduct exploration there. Earlier, we had recognized that the favorable precious-metals terrain of Southern California and New Mexico probably continued southward into Mexico, and we wanted to start a gold rush there similar to what we were seeing in California. So in late 1983, I had written an exploration proposal and submitted it to the Mexican mining company Compañia Minera Las Cuevas, S.A., of which Noranda owned 49 percent. This is the company that owned the Las Cuevas fluorspar mine that I had visited in 1982. In my proposal, I recommended that Noranda and Las Cuevas work together to explore the gold potential in the northern part of the Mexican states of Sonora and Chihuahua, right at the US border. Compañia Minera Las Cuevas liked my proposal and wholeheartedly supported it.

So in the spring of 1984—with new visas in hand—Jim Norris and I accompanied Arturo Geyne, the managing director of Compañia Mineral Las Cuevas, in flying in a fixed-wing aircraft over northern Mexico to become familiar with the topography, geography, and geology. Another great adventure was in the offing. Unfortunately, one morning it was pretty rough flying because of highly turbulent air, and I became airsick. This rarely happened to me.

Following our aerial flyover, we conducted field reconnaissance on the ground in northern Sonora and northwestern Chihuahua where we were successful in identifying attractive gold prospects at Quitovac and El Antimonio. We recommended to Compañia Mineral Las Cuevas that they acquire and drill both of them. However, I never did learn the fate of these two prospects.

That part of Mexico along the border with the United States is remote and at that time was quiet and almost devoid of people. But with the dangers of drug running and illegal immigration along that border today, I don't think it would be safe now to do what we did in 1984. In fact, I don't know if one would even be allowed to fly over that area in a low-flying airplane.

Back in the fall of 1983, Jim Norris had become interested in the Bagdad-Chase gold property in the Mojave Desert south of Ludlow, California. So Jim, Bob Page, and I went to look at it one day and found remnants of a typical late 1800s-early 1900s mining camp that was dotted with old mines, their headframes (a wooden or metal structure designed to lower a cage or elevator into a mine shaft), and a few rundown buildings. There once was a town there named Stedman, but little of it remains. It does not even show up on a highway map. Over the years, I saw many of these old ghost towns scattered throughout the west. From there, we went to Nipton, California, to look at another gold prospect in the New York Mountains near Las Vegas.

The next day, we prospected around the old Doble mine in the San Bernardino Mountains and at the Dinner Bell property in the adjacent San Gabriel Mountains. These two mountain ranges constitute the northern boundary of the Los Angeles metropolitan area, and they limit the northern growth of the city. We needed to check the claim status in these two ranges to see if there had been any recent claim-staking activity by our competitors. This required a trip to the San Bernardino County courthouse in the

City of San Bernardino (population 205,000), one of the major cities of this huge mass of humanity.

San Bernardino is a large county, the largest in the continental United States and larger in area than the state of Rhode Island. Going to San Bernardino was unusual for us since county courthouses that we usually visited were in little places like Dillon, Montana (population 3,700), the county seat for Beaverhead County; or Westcliffe, Colorado (population four thousand), the county seat for Custer County. On many of our days there in Southern California, we couldn't see the tops of the mountains from the valley below because of thick, ugly, brown smog. This was not a good place to conduct exploration geology, I thought; too many people and bad air. Nevertheless, going to that courthouse was a necessity.

We were encouraged by the rocks, however, so we went back later on a follow-up trip to look at other prospects in the San Gabriel and San Bernardino mountains. This time we had a good view of the San Andreas fault, where it separates these two mountain ranges at El Cajon Pass. The San Andreas fault, extending nearly eight hundred miles from near the Mexican border to north of San Francisco, is one of the world's major tectonic plate boundaries, called a transform fault. It forms the boundary between the Pacific plate on the west and the North American plate on the east. Displacement on the fault is mostly horizontal, and the Pacific plate, including the Los Angeles metropolitan area, is moving north relative to the rest of North America. Geologists refer to this as right-lateral strike-slip fault movement. You can think of the earthquakes produced along the fault as the incremental steps of tectonic plate movement.

The displacement rate on the fault is about two inches per year, fast enough to move rocks about one mile in sixty thousand years. On a geologic scale, movement on the San Andreas fault is frequent enough to be considered continuous. Displacement along the fault began about 15–20 million years ago, and since then, Los Angeles has moved north on its way to Alaska. If we can wait around long enough (another fifteen million years), Los Angeles will be next door to San Francisco.

Just north of Los Angeles, the San Andreas fault has cut off the southern end of the Sierra Nevada mountain range and transported it north to around San Luis Obispo on the central coast. If you look on Google Earth, you can clearly see the straight-line expression of the San Andreas fault and put your finger right on it. For years, this fault has been poked and prodded and monitored by the US Geological Survey in an attempt to learn how to predict earthquakes. Alas, progress has been made, but all we can say at this point is that another major earthquake will occur, but we can't say when.

After returning to Tucson, I was summoned to Denver for a day's talk with Geoff Snow, Noranda Exploration's president, about the future of the Tucson office. He wanted to know what we were finding and what I thought the likelihood was of our making a discovery. The changing and uncertain economic environment facing mining companies was calling for some drastic measures on Noranda's part. Geoff said he would get back to me on what he decided; this did nothing to relieve my anxiety. Things were not looking good.

However, fieldwork must go on, so in mid-April Bob Page and I drove to Ludlow, California, where we met Jim Norris and took another look at the Bagdad-Chase property. We saw favorable geology with good breccias (broken rocks that sometimes have open spaces filled with precious metals). Gold was discovered there in 1898 by a railroad roadmaster who was out searching for water. Mining continued into the mid-1900s, and at one time the Bagdad-Chase Mine was the largest gold producer in San Bernardino County. We were unable to acquire the property, however.

Since drilling was now finally underway at the Winkler Anticline in New Mexico, I took a quick trip there to see what Jim and Ellie were finding. I spent the night in the trailer camp they had established on the property to serve as their home base. Tents were set up to handle the overflow when they had company—that is, visiting geologists. The nearest town was Lordsburg, New Mexico, a town that I never cared for, and the trailer and tents provided much better accommodations anyway. Jim and Ellie were doing a superb job of logging drill core and managing the project, and their drilling program was intersecting attractive-looking rocks.

For some reason forgotten by me, a few days later I was back in my old stomping grounds of Montana looking at the Dreadnaught and York properties near Helena. Noranda's Northwest office probably brought me back to consult with them. It was always nice to feel needed.

Back in the Southwest, I drove to Las Vegas with other geologists from Noranda's Tucson office to look at the Morning Star and Roadside properties just across the state line in Southern California. The next day, we stood on the Plomosa detachment fault and the Mule Mountain thrust fault in the Mule Mountains. Detachment faults are unusual because they are nearly horizontal; you don't commonly get to stand on a fault, so this was a treat.

Then, the hammer dropped. On June 11, Geoff came to town to inform us that Noranda's southwest district office would close on August 15, and all personnel and equipment would be transferred to the Reno office. Noranda's US exploration effort was being consolidated to reduce costs.

When looking back on it now, the mid- to late-1970s were probably the golden years of modern minerals exploration in the United States. Many companies were in the exploration business, jobs for geologists were plentiful, discoveries were being made, and new mines were being developed. By 1984, however, poor worldwide economic conditions and slumping metals prices had forced many companies out of the exploration business, and the future of exploration in the United States did not look promising. Between 1981 and 1987, US exploration expenditures declined by 73 percent. Within the exploration community, the number of exploration professionals employed by US companies declined from 2,355 in 1981 to 1,277 in 1984. As it turns out, the slump lasted until about 1988, and exploration in the United States never did fully recover. The decline in minerals exploration was caused by low metals prices and excess worldwide supply. Ironically, we exploration geologists had been too successful.

Nevertheless, life must go on, so after hearing that news, and after counseling my staff that everything would work out fine for them—which it did—I went back to looking at gold prospects and operating gold mines in southeastern California near the Colorado River: Gold Fields' Mesquite gold mine, the Picacho mine again, and the old American Girl mine in the Cargo Muchacho Mountains east of Brawley. At the Picacho mine, I have pictures of three of us on our hands and knees examining the flat surface of the detachment fault that controls the location of the ore. As it turns out, this part of California was indeed favorable for exploration because by the 1990s there were six operating open-pit gold mines there.

Our earlier, cursory examination of rocks at Tumco had been sufficiently encouraging that we decided to give it a more thorough evaluation. I assembled a crew of half a dozen Noranda geologists from other districts, and we spent six days geologically mapping and sampling in and around this old gold mine in the Cargo Muchacho Mountains. Javier Contreras, our draftsman in the Tucson office, and his wife, Donna, assisted the field crew by serving as our cooks and camp tenders. Skip Svendsen, our always-helpful technician, had the toughest job—he hauled out all the rock samples we collected for assay.

Performing such work in late June, early July, two of the hottest months of the year in the Sonoran Desert, was probably not my best idea. Boy was it hot! One day, a thermometer lying by one of our trucks read 135°F—that was anomalously high because of heat radiation from the vehicle. Even though our alarm clocks went off at 4:30 a.m. so we could try to beat the heat, temperatures reaching 115°F in the afternoon were common. I did notice, however, that one of our geologists did find enough spit to whistle a tune during one of his traverses, and by the end of our mapping and sampling stint, we built campfires to cool off by. This is a reflection of the spirit and stamina of geologists that we were able to endure this assignment in good humor.

What a deplorable working environment this must have been for the miners laboring away here in the early 1900s. The desert in general, and at Tumco in particular, is a stark place, especially at that time of year. Nothing was present except the black, desert-varnished rocks, the sky, the debilitating and blazing sun, and the ten of us. The mountains, barren of vegetation and soil, were like the moon, just hotter and with more gravity. Thank goodness for good meals and cold beers provided by Javier and Donna.

On July 5, I flew to the mining camp of Timmons, Ontario, home of Noranda's Pamour-Porcupine gold mine. There I talked with mining personnel about a possible job for me as mine geologist, but I was not intrigued about moving there. So on August 17, 1984, I submitted my resignation from Noranda Exploration, Inc., after almost ten years with the company. August 24 was my last day of work. The rest of the staff of the Tucson office moved to Reno and continued to conduct exploration from there. For me to move to Reno would have meant a demotion, and although a spot could have been found for me elsewhere in Noranda, it was time to move on, to see what adventures would come in the next chapters of my life.

EPILOGUE

I am reminded of that famous book *Oh, the Places You'll Go,* written by Dartmouth alumnus "Dr. Seuss," class of 1925. My minerals exploration career took me to places I had never dreamed of seeing and, in many cases, places I had never even heard of. When I was at Dartmouth and decided to major in geology, it never occurred to me the extent of adventures I would have, the places I would see, the science I would study, and the fascinating people I would come to know. And I never expected that through my work I would make a contribution, however slight, to the country's economic well-being. I'm happy to say that my love of geology, traveling, and adventure continues to reward me today. I never regretted taking the road less traveled.

Geology gives birth to lifelong investigation and puzzle solving, if you have the curiosity, the know-how, and the time and inclination to pursue it. Because of my love of travel and adventure and my exploration experience, I have been fortunate to have had the luxury of being able to indulge my lifelong passion for geology. With members of my family, I have visited Devils Postpile in northern California, one of the best examples of columnar basalt in the world; Wallace Creek in central California, where 420 feet of offset along the San Andreas fault is superbly obvious; Meteor Crater in northern Arizona, clearly the best example in the world of a crater formed by a meteor's impact; fjords cut into Norway's coast by massive glaciers; tubes in lava at Newberry National Volcanic Monument in Oregon through which lava once flowed. These are just a few examples.

After leaving Noranda Exploration, I became a consulting exploration geologist for a few months, and then my family and I moved to the San Francisco Bay Area. There I worked for an environmental engineering firm before moving on to Lawrence Livermore National Laboratory in Livermore, California. Over a twenty-year period, I managed the groundwater cleanup program there. These were productive and satisfying jobs, but they did not provide the same excitement and adventure that exploration geology had.

As the famous mid-1900s photographer Diane Arbus said: "My favourite thing is to go where I've never been." Going places where I have never been before has provided a fun-filled, adventurous life that continues to this day.

GLOSSARY OF TECHNICAL TERMS

Adit   a man-made horizontal passage dug from the ground surface for exploration or mining

Anticline   an upward fold in sedimentary rocks

Archean   the oldest part of the Precambrian eon, 2.5 billion years old and older

Arête   a narrow, jagged mountain ridge sculpted by glaciers

Arroyo   a desert gully

Ash   fine-grained particulate material erupted from a volcano

Assay   to test ores or minerals for purposes of determining the amount of valuable metals they contain

Azurite   a deep-blue copper carbonate mineral

Badlands   barren, rough, and gullied topography in arid areas

Basalt   fine-grained, dark- or medium-colored rock produced by a volcanic eruption; the most common extrusive igneous rock

Base metals   nonmagnetic metals such as copper, lead, zinc

Basement   undifferentiated rocks of igneous or metamorphic origin that underlie the oldest sedimentary or volcanic rocks

Bastnaesite   a carbonate-fluoride ore mineral containing rare-earth elements

Batholith	a large, intrusive igneous mass of more than forty square miles in surface exposure—for example, the Sierra Nevada batholith
Bedrock	a general term for the solid rock exposed at the Earth's surface or that underlies unconsolidated surficial materials such as soil
Bomb	a mass of volcanic rock that was semisolid when erupted from a volcano
Borax	a sodium tetraborate or disodium tetraborate mineral mined from evaporite deposits; a component of many detergents, cosmetics, and enamel glazes
Brachiopod	a bivalve marine invertebrate resembling a clam; lived from the Early Cambrian period to the present
Breccia	a rock consisting of angular rock fragments in a finer-grained matrix
Calcareous	consisting of calcium carbonate
Caldera	a large, generally circular volcanic depression caused by the collapse of the roof of a magma chamber following eruption of magma
Carbonate	a combination of carbon and oxygen, often combined with calcium to form the mineral calcite
Carbonatite	an intrusive or extrusive igneous rock consisting of greater than 50 percent carbonate minerals
Chalcocite	a copper sulfide mineral (Cu_2S); an ore mineral of copper
Cirque	a large, amphitheater-shaped basin carved into the side of a mountain by a glacier
Colemanite	a mineral composed of calcium, boron, oxygen and water ($Ca_2B_6O_{11} \cdot 5(H_2O)$); a source of boron

Crater	a bowl-shaped, rimmed feature forming the summit of a volcano
Crinoid	a flowerlike marine invertebrate, usually attached by a "stem" to the ocean floor; in fossils, the stem is what is usually preserved; lived from the Ordovician period to the present
Crop out	to be exposed at the ground surface
Deposit	a naturally occurring body of rock containing a concentration of valuable minerals
Desert varnish	a shiny, brown and black chemical varnish on rock faces composed of clay and iron and manganese oxides
Detachment zone fault	a low-angle fault that separates Tertiary rocks from crystalline basement rocks
Dike	a tabular, igneous intrusion that cuts across older surrounding rocks
Drift	a man-made horizontal passage within a mine; often follows a vein
Escarpment	a long cliff or steep slope separating two comparatively level areas
Exploration	the search for mineral deposits and the work done to establish their extent
Extrusive (volcanic) rocks	igneous rock that has been ejected or erupted onto the surface of the Earth, such as rhyolite tuffs and basalt flows
Fault	a fracture or break in rocks along which movement has occurred
Fault scarp	a steep slope or cliff representing the exposed surface of a fault
Fluorite	a mineral consisting of calcium and fluorine (CaF_2)
Fluorspar	the commercial name for fluorite
Fumarole	an opening in the Earth's crust, often on or near a volcano, that emits steam and gases

Glacial drift or till	a general term for rock material transported and deposited by a glacier
Gneiss	pronounced "nice," a coarse-grained metamorphic rock consisting of bands of alternating zones of light and dark minerals
Gossan	a cellular accumulation of iron oxide material derived from leaching of sulfide minerals; usually found at or near the ground surface
Granite	a light-colored, coarse-grained, intrusive igneous rock; rhyolite is the extrusive equivalent
Headframe	a wooden or metal structure designed to lower a cage (elevator) into a mine shaft
Hematite	the principal ore mineral of iron (Fe_2O_3)
Huebnerite	a red to black manganese-tungstate mineral, ($MnWO_4$); an ore of tungsten
Hydrothermal alteration	modification of rocks or minerals by hydrothermal fluids
Hydrothermal fluid	hot, subsurface mixture of water, gasses, and dissolved materials
Igneous	a class of rock that solidified from molten magma, either intrusive or extrusive
Intrusion or intrusive rock	a mass of igneous rock that intruded preexisting rock
Jasperoid	a rock, commonly limestone, whose minerals have been replaced by silica
Joint	a fracture in rocks along which no appreciable movement has occurred
Kimberlite	a dark-colored, heavy rock rich in mica, garnet and diopside that may contain diamonds
Laccolith	an igneous intrusion having a flat floor and a domed roof

Lava	a general term for magma or molten rock that has flowed onto the Earth's surface; also refers to the rock that solidified from it
Lithified	the process by which unconsolidated sediment is converted into rock
Loess	wind-blown glacial silt; rhymes with "puss"
Mafic	containing large amounts of iron and/or magnesium
Magma	molten rock produced within the Earth, from which intrusive and extrusive igneous rocks form
Malachite	a deep-green copper carbonate mineral, $Cu_2(CO_3)(OH)$
Massive sulfide deposit	an assemblage of metal sulfide minerals (mainly Cu-Pb-Zn) that is associated with and created by volcanic-associated hydrothermal events in submarine environments
Metamorphic core complex	a domelike mountain cored by deep crustal rocks
Molybdenite	a steely gray mineral consisting of molybdenum and sulfur (MoS_2)
Molybdenum	an element (Mo) used in alloy steel
Niobium	a metal (Nb) that improves the strength of steel; sometimes called columbium
Normal fault	a break in rocks in which one fault block has moved down relative to the rocks on the other side of the fault
Obsidian	dark, volcanic glass that fractures in a conchoidal pattern
Ophiolite	a remnant of oceanic crust that has been thrust upon continental crust by plate tectonic movement; it represents a tectonic plate boundary

Ore	rock containing valuable minerals that can be mined at a profit
Orogeny	a period of mountain building
Outcrop	an exposure of bedrock at the Earth's surface
Overburden	valueless rock and unconsolidated materials lying above an ore deposit
Overthrust	the upper plate of a fault where movement has mostly been horizontal
Phenocryst	a relatively large, conspicuous crystal in a matrix of smaller crystals of an igneous rock
Plate tectonics	the movement and interaction of the rigid plates composing the Earth's outermost layer
Plateau	a large, relatively high area bounded on one or more sides by cliffs or steep slopes
Playa	a dry, vegetation-free flat area at the lowest part of an undrained desert basin
Pluton	a general term for an igneous intrusion; think of it as a fossilized magma chamber
Porphyry	an igneous rock that contains conspicuous phenocrysts in a finer-grained groundmass
Portal	the surface opening to the underground workings of a mine
Precious metals	any of several metals, including gold, silver, platinum, and palladium, that have high economic value
Prospect	a potential site of a mineral deposit or an area that has been explored in a preliminary fashion by digging
Prospecting	the application of geologic, geochemical, or geophysical techniques to discover mineral deposits
Proterozoic	the youngest part of the Precambrian eon, 542 million years to 2.5 billion years old

Pyrochlore a yellowish to reddish mineral composed of niobium, oxygen, and fluorine, $(Na,Ca)_2Nb_2O_6(OH,F)$

Quartz a mineral composed of silica and oxygen, (SiO_2)

Quartz monzonite a light-colored intrusive rock similar to granite

Rare-earth elements the seventeen metallic elements on the periodic table whose atomic numbers are 57 through 71; so named because they were first separated from rare minerals

Rhodochrosite a manganese carbonate mineral $(MnCO_3)$, often found as cherry-red crystals

Rhyolite a fine-grained, light-colored igneous rock; the extrusive equivalent of granite

Rift a deeply faulted valley where two parts of the Earth's crust are separating

Scheelite a yellowish-brown ore mineral of tungsten, $(CaWO_4)$

Schist a medium-grained to coarse-grained metamorphic rock composed of laminated, often flaky parallel layers of micaceous minerals

Shaft a man-made vertical passage dug from the ground surface for exploration or mining

Silicified, silicification impregnated with silica, such as rock or fossil wood that has been replaced with silica

Skarn a variably colored green or red metamorphic rock formed where igneous intrusions come in contact with calcareous rocks

Soda ash a significant economic commodity produced from trona; it has applications in manufacturing of glass, chemicals, paper, detergents, textiles, and food. It is an ingredient in both sodium bicarbonate (baking soda) and sodium phosphate (detergents)

Stock	an igneous intrusion that is less than forty square miles in surface exposure
Stockwork	a crisscrossing network of veins
Strata	tabular layers of sedimentary rock
Stratovolcano	a cone-shaped volcano built of alternating layers of lava and volcanic ash
Strike-slip fault	a break in rocks in which one fault block moves horizontally relative to rocks on the other side; the San Andreas fault is an example
Stromatolite	a colony of blue-green algae that builds mounds resembling cabbage heads
Tarn	a small mountain lake formed in a cirque excavated by a glacier
Tectonic plate	a part of the Earth's crust that moves in response to forces in the underlying core; nine major plates and several minor plates are recognized, though the actual number is in debate
Terrane	a body of rock bounded by faults and characterized by a geologic history different from adjacent terranes
Trilobite	an extinct marine arthropod characterized by a three-lobed exterior skeleton
Trona	a sodium carbonate mineral, $[Na_3(CO_3)(HCO_3) \cdot 2(H_2O)]$, mined from evaporite deposits
Tuff	a rock composed of compacted and cemented volcanic ash
Ultramafic rock	a dark-colored igneous rock comprised primarily of iron and manganese minerals
Varve	a layer of fine-grained sediment deposited in a body of still water during the course of one year

Vein	a thin, sheet-like body of minerals that has precipitated from hydrothermal solutions in cracks and crevasses of rocks
Volcanic neck or throat	an erosional remnant of volcanic rock that formerly filled a volcano's conduit
Workings (mine)	a general term that refers to any excavation made underground, such as a tunnel, shaft, adit, etc.

Printed in the United States
By Bookmasters